# ASTROLOGY IS
# QUANTUM ENTANGLEMENT

# ASTROLOGY IS
# QUANTUM ENTANGLEMENT

## ASTRO QUANTUM ENTANGLEMENT

Vikram Divakar & Soundar Divakar

White Falcon Publishing

Astrology is Quantum Entanglement
Vikram Divakar, Soundar Divakar

Published by White Falcon Publishing
Chandigarh, India



ISBN - 979-8-89222-349-2

Dedicated To

**Aishu & Sasi**

# Contents


Preface ............................................................................... ix

**CHAPTER 1** Implications of Astrology ........................................ 1

**CHAPTER 2** Distinct Features of Astrology ............................... 5

**CHAPTER 3** Newton's Gravity and Einstein's Space-Time Distortion ...... 73

**CHAPTER 4** Probable Concepts in Physics to Explain Astrology .......... 84

**CHAPTER 5** Astrology is Quantum Entanglement ................................ 92

**CHAPTER 6** Inherent Problems in Astrology ........................................ 113

References ........................................................................... 128

# Preface

Earlier, in our book on Astrology – A Science of the Quantum World, we have arrived at the conclusion, that astrology is more closely related to Quantum mechanical concepts other than any other branches of physics. Thus we explored Heisenberg's Uncertainity principle, Quantum Entanglement, Quantum Gravity, General and Special Theory of Relativity of Einstein in relation to Space-Time distortion, String-Super String-Brane-M Theories, Dark Energy and Lensing Effects of Planets, as possible concepts towards understanding the influence of planets on the genetic materials of human beings.

Astrology – A Science of the Quantum World, Vikram Divakar and Soundar Divakar, 2021, Notion Press, Chennai, India. (Ref 1)

In this present book, we attempt to explain why Quantum Entanglement alone could explain, most, if not all the subtle aspects of astrology. Futher the Noble prize in Physics awarded to the subject of Quantum Entanglement in 2022 served as a boost to explain astrology on scientific terms. In astrology, this concept of Quantum Entanglement, has been taken further, as explained in this book, to deal with various aspects of hidden properties of human beings, governed by human DNA molecules, to be influenced by external factors prevalent in the space through the planets, over large distances.

This book is the culmination of 50 years of research on the SCIENCE behind Astrology. The astrological part of this book deals with mostly Naadi System of Prediction in relating astrology to science. The Naadi system is an ancient system of prediction which is known for its accuracy than even other Indian systems like Parasari and Jaimini systems. This system is more than 2500 years old. The rules are written in palm leaves and has been maintained as close guarded secret by different schools of astrology and there are quite a few of them. Due to unstinted efforts of the authors, most of these rules have been published in the following book:

1. Language of Rahu (Naadi Rules Explained), by Vikram Divakar and Soundar Divakar, 2016, published by Notion Press in India. Available on Amazon.in and Flipkart.com. (Ref. 2)
2. Rahu Naadi, Vikram Divakar, and Soundar Divakar, 2024, White Falcon Publishing, Chandigarh, India (Ref 3).

Planetary longitudes employed in this book, were obtained from NASA Epimehris, and has been subjected to Precessional correction of the Indian noted astrologer, Bangalore Venkata Raman, for the year 386 ACE, considering a precession of 51" every year. Indian astrology does not consider precessional effect, i.e., the demarcation of signs have not been corrected since 386 ACE, unlike Western Astrology. Probably, that could be a reason for obtaining better accuracy in the Indian method compared to the Western system.

# Chapter 1

# Implications of Astrology

Before we proceed to explore the science in astrology, we should understand what astrology implies.

Astrology is applicable to all the people of this world. Horoscopes do not differentiate between religion or class or caste or creed or nationality. Planetary positions of people born under similar planetary configuration are the same. The difference is manifested only in the Genetic-Make-Up (GMU) and the environment, both of which can alter the outcome to a significant extent. Planetary effects on the GMU of the individual are more important for a particular planetary configuration to manifest its full effect on the individual. In several horoscopes mere planetary configuration do not give even near-to reasonable predictions. This is because the planetary influence and the GMU for the individual do not go hand in hand.

**Planetary configuration and GMUare the two main factors, why, even individuals coming from the lower rungs of the social ladder outshine others from higher levels of the society, most of the times. What all is required is a matching GMU responding to the effects of the planetary configuration at the time of birth. The environmental**

**effect plays a very small, but an important role. Environmental effect per-se, cannot and does not change the outcome to a great deal. This is the reason why several Scientists, Politicians, Artists, Sports personalities, Journalists, Religious personalities, Writers, Authors and other prominent personalities emerge from lower rungs of the society from time to time. The 'Chosen people' with 'Blue blood' from upper classes, especially those with 'legacies', do not lead a prominent life for the same reason. In most of these 'upper crust' cases, the individuals may not have a matching horoscope with their Genetic-Make-up, in spite of being born in an affluent environment. Thus, planetary configuration lays down uniform yardstick for all the individuals, only to vary due to GMU to a large extent and Environmental effect to a lesser extent.**

Another aspect of astrology is that there is no right or wrong in astrology. At the most, it tries to describe what an individual is capable of performing. Similarly the moral and ethical issues of evil and good or right or wrong or honest and corrupt do not find a place in astrology. Astrology implies that these moral and ethical issues are like two sides of a coin, that exists as a matter of fact. Astrology does not also justify evil or bad. Astrology presents them as Distraction or Obstruction factors that are bound to happen with the appropriate disposition of the Genetic Make-up. Astrology also indicates the punishment for an 'evil-doer'.

Several people are born all over the world under identical planetary configurations that exist for a duration of a few minutes to a few hours. While one person goes on to become a Sociopath, Psychopath, or a Serial Killer, another person with the same planetary configuration could be a harmless individual. Thus, astrology tries to imply that we are living in more or less a PERFECT WORLD. The differences are due to the variances in our Genetic Makeup, bolstered by the self-imposed constraints thruston us by our own SOCIAL STRUCTURE, SOCIETAL NORMS, SOCIAL PRACTICES, and SOCIAL PERCEPTIONS that have evolved over thousands of years. The important truth that astrology drives home is that our SOCIETY IS NOT FULLY EVOLVED and IT IS STILL EVOLVING.

Planets cannot be blamed for somebody's actions. The perpetrator has to take the responsibility for his/her actions. The same planetary combinations, present in individuals born at the same time do not produce sociopaths or psychopaths or angels or saints or leaders. That is dictated by the GMU that alters the outcome to a great deal. This is not also the place and time for dwelling into ethical or moral values of actions of individuals. A clear understanding of the entire scenario, from that of the perpetrator to the victims and their relatives, the environmental set up available to both the perpetrator and the victim and the accepted social conditions prevalent at that time, all come into play to decide a decision on this complex matter. THE BLAME SHOULD NOT REST ON MERE PLANETARY CONFIGURATIONS ALONE.

We should be constantly examining our way of life and apply mid-course corrections. Unfortunately, the successful people in our society, do not allow that to happen. That is the reason why major calamities like invasions, migrations occur from time time to act as forced mid-course events which eventually brings out some sort of required change, though not fully. From time to time, individuals evolve with monstrous qualities like the ruthless invaders and dictators who are probably required to bring out these changes.

Another important message conveyed by astrology is that, it does not pronounce that successful people are correct, although their words and actions are lapped by a vast majority of the Borchevaistic population constituting to about 70% or so, who are not that successful in life.

In fact, astrologically, the 70% of our population that are not successful are those, who possesses a GMU which do not match with their planetary configuration at the time of their birth probably without any exception.

If Karl Marx (Ref 4) had understood astrology, his books would have conveyed a different message. Astrology also emphasizes that all are not equal at anytime. There is always some difference between each and every individual. To a very large extent, their planetary configuration are not the same. However, in a very small percentage of cases, where the planetary configurations are the same, their GMU differs distinctly. This is a SYMMETRY problem in Physics where, SYMMETRY LAWS do not replicate another similar entity at any time. As per symmetry laws, no two dimensional entities are identical. This applies to human beings as well. Even genetic cloning may not produce identical human beings, like what is depicted in movies. It is highly possible that it boils down to the fact that nobody is equal to the other at any time. Planetary configuration in conjunction with GMU brings out this difference distinctly.

Chapter 2

# Distinct Features of Astrology

The ecliptic path followed by the planets around the Sun is divided into 12 regions or signs of the zodiac as they are normally called. Each sign is associated with one planet, who is the lord of the sign, one or more planets termed friends, one or more planets designated as enemies, and a few as neutrals, with few signs even having an exaltation planet and debilitation planet. Although it is a misnomer to call the Sun and Moon planets, they are considered essential celestial bodies that influence human lives. By planets, Sun, Moon, Mercury, Mars, Jupiter, Saturn and the two nodes of Moon (North Node and South Node called Rahu and Ketu respectively in Indian texts) are considered essential for such a study.

Other outer Saturnian planets like Uranus, Neptune, and Pluto (although it has lost its planetary status) are mainly used by Western astrologers. This book, which mainly deals with the Indian system, does not consider those outer Saturnian planets to be significant enough to influence human lives. The authors are also not sure whether such an exclusion of these outer Saturnian planets is rational or not. From their extensive analysis of several horoscopes, they have arrived at the conclusion to ignore them as their effect was deduced to be insignificant.

According to the Indian system, the background space of each sign starting with Aswini 0° as the beginning of Aries is important. Thus Indian systems have inadvertantly attributed lot of influence to the background space as responsible for influencing humans which formed the basis of 27 constellations, each of 13.33° and division of each constellation into four parts of 3.33° and two-and-a-quarter constellation to a Zodiacal Sign of 30° each and so on.

The signs and positions of planets at the time and date of birth of the individual can be obtained from any standard horoscope programs. Astrology explains various aspects of the individual's life such as Circumstances of Birth, Physical and behavioral attributes of the person, Description of Father, Mother, Brother, Sisters, Wife/Husband, Friends, Companions (lovers if any), Marriage (one or many), Married Life, Children, Education, Profession (Nature and Success therein), Finances, Affluence (Period for Possession of house, Vehicles, wealth if indicated by Planets), Enemies, Loan, Diseases, Losses, Gains, Longevity, General prosperity, and so on through empirical correlations. Nature and periods of occurrences of these aspects are described by astrology. Astrology, over a period of time, has come up with quite a lot of empirical correlations, relating the configuration of planets at the time of birth to the life and events of an individual. Several time-tested combinations are thus available in astrology, arrived at through these empirical correlations, although some are not rigorously and statistically tested. Even Astrologers have not taken it upon themselves to test such combinations statistically. The reason for this is the conflict in interest exhibited by the astrologers all over the world, as they never came together to show the world a single predictable astrological method, with merits and demerits, which has only led to the downfall of this important subject through the ages, unable to withstand the onslaught of Scientific rationality and logistics.

There are several important scientific concepts dealt with by astrology.

1.  Astrology always works in relation to the background Universe. Why should Astrology bother about the background scenario of the ecliptic path? The universe, in its three-dimensional nature is spread all around our solar system and our Galaxy. The ecliptic background comprises, humpty number of Galaxies with multitude of stars and gaseous Nebulae, Neutron stars, Pulsars, Quasars, Black holes among other things. Thus, the whole Universe is full of Energy, the Signs(or Rasis) depicts variation in this Cosmic Energy over the entire ecliptic path which the planets follow. We now know of dark-matter and dark energy also, that we cannot see, besides usual matter and energy, which we can see and infer. Background celestial objects play a crucial role in astrology in not only defining the boundaries of zodiacal signs, but also in defining attributes describing an individual. In the backdrop of the zodiacal background, each planet exhibits its own attributes, some more intensely than others and few malefically as well. Following figure illustrates format of the charts used in this book with the 12 zodiacal signs shown. Although there are several ways to illustrate them, authors have preferred this mode since they are familiar with it.

| PISCES | ARIES | TAURUS | GEMINI |
|---|---|---|---|
| AQUARIUS | **SIGNS** | | CANCER |
| CAPRICORN | | | LEO |
| SAGITTARIUS | SCORPIO | LIBRA | VIRGO |

| Jupiter | Mars | Venus | Mercury |
|---|---|---|---|
| Saturn | LORDSHIPS | | Moon |
| Saturn | | | Sun |
| Jupiter | Mars | Venus | Mercury |

2.  In astrology, human beings are described as multi-dimensional entities, with visible and hidden dimensions. Astrology deals with each individual as a dimensional entity and describes all the attributes of the individual. It talks about appearance, health, wealth, intelligence, relations, family, wife/husband, friends, longevity, diseases, honour, profession, luck and even losses. While, some of these aspects are visibly noticeable, others involving potentialities like intelligence, luck, ability to earn wealth, ability to acquire skills, description of an individuals family (mother, father, brother, sisters, relatives, wife, children) are invisible and hidden. Thus, here is an area where an individual is described by his/her multi-dimensional attributes - both visible and hidden. **Unifield**Field Theory and its related **String Theory** are still contemplating on 10-32 dimensional (or more) universe with four dimensions of space and time and the remaining probable hidden dimensions. Astrology clearly implies the multi-dimensional attributes of the universe.

The following Tables illustrate these aspects.

# Table 1

## Lordship, Friendship, Enmity, Neuralities, Exaltation and Debilitation of Planets

### 1. Aries

| Angular Coverage | 0-30° |
|---|---|
| Lordship | Mars |
| Exaltation | Sun |
| Debilitation | Saturn |
| Direction | East |
| Element | Fire |
| Friends | Sun, Moon, Jupiter, Ketu |
| Enemies | Mercury, Saturn, Rahu, Venus |
| Neutral | None |
| Attributes | Energy, Lands and Land Produce, Agriculture/Cultivation, Head, Short-Tempered, Egoistic, Materialistic, Practical, Harsh words, Out Spoken, Active, Army, Police, Quarrelsome, Aggressive, Hasty and Impatient, Obedience to elders, Rigid Personality, Boastful, Liking for Fame, Recognition and respect in Society, Ruthless, Sadness, Discriminative at times |

## 2. Taurus

| | |
|---|---|
| **Angular Coverage** | 31-60° |
| **Lordship** | Venus |
| **Exaltation** | Moon |
| **Debilitation** | None |
| **Direction** | South |
| **Element** | Land |
| **Friends** | Mercury, Saturn, Rahu |
| **Enemies** | Sun, Jupiter, Mars, Moon |
| **Neutral** | Ketu |
| **Attributes** | Artistic, Appearance conscious, Pleasant, Pleasing Manners, Boastful, Jewelry, Precious stones, Finances, Commerce, Trade, Attractive, Lavish Spending, Attraction for the opposite sex, Determined, Strive to achieve the goal, Silk, Fashion Clothes, Women ware, Proficiency in music, Fine arts, Cine field, Houses, Vehicles, Drama, Show business, Treasury, Pleasure loving |

## 3. Gemini

| | |
|---|---|
| **Angular Coverage** | 61-90° |
| **Lordship** | Mercury |
| **Exaltation** | None |
| **Debilitation** | None |
| **Direction** | West |
| **Element** | Air |
| **Friends** | Sun, Venus, Saturn, Rahu |
| **Enemies** | Jupiter, Mars, Moon |
| **Neutral** | Ketu |
| **Attributes** | Skills and Skill oriented Jobs, Creative intelligence, Good Memory, Educational Institutions, Tailoring, Cloth Manufacture, Giving shapes, Computers, Mathematical Proficiency, Liking for Travel, Proficiency in music, Sculptor, Painter, Highly Analytical, Knowledgeable, Speech, Landed Property, Research, Communication, Intellectual Manipulation, Good understanding of Mass psychology |

## 4. Cancer

| | |
|---|---|
| **Angular Coverage** | 91-120° |
| **Lordship** | Moon |
| **Exaltation** | Jupiter |
| **Debilitation** | Mars |
| **Direction** | North |
| **Element** | Water |
| **Friends** | Sun, Mars, Ketu, |
| **Enemies** | Saturn, Mercury, Venus, Rahu |
| **Neutral** | None |
| **Attributes** | Wavery, Changing, Emotional, Travels, Wandering, Eatables, Food, Agricultural products, Cattle products, Cereals and Grains, Drinks, Chemicals, Army, Police, Female folk, Intelligent, Knowledgeable, Foreign connections, Prosperous, Change of Place, Foreign Residence, Overseas Travel, Flowers, Arts and Arts related, Cheat, Deceit, Thief, Elopement, Foreign travel, Travel overseas, Desertion by wife, sisters, friends (opposite sex), losing people in one's life. |

## 5. Leo

| | |
|---|---|
| **Angular Coverage** | 121-150° |
| **Lordship** | Sun |
| **Exaltation** | None |
| **Debilitation** | None |
| **Direction** | East |
| **Element** | Fire |
| **Friends** | Mars, Moon, Jupiter, Ketu |
| **Enemies** | Saturn, Venus, Rahu |
| **Neutral** | None |
| **Attributes** | Fire, Chemicals, Government service, High post, Rigid person, Position of Authority, Discriminative, Practical thinking, Fast decision making, Egoistic, Boastful, Materialistic, Practical, Favors from Rulers and Royalty, Political Connection, Foreign Travel, Highly placed, Royal and Royalty Connections, Medicine, Drugs, Doctor, Surgeon, Fame, Haughty, Forceful, Play with emotions, Success in endeavours, Name, Fame, Command in Office and Society, Extraordinary Determination. |

## 6. Virgo

| Angular Coverage | 151-180° |
|---|---|
| **Lordship** | Mercury |
| **Exaltation** | Mercury |
| **Debilitation** | Venus |
| **Direction** | South |
| **Element** | Land |
| **Friends** | Sun, Saturn, Venus, Rahu |
| **Enemies** | Mars, Moon, Jupiter |
| **Neutral** | Ketu |
| **Attributes** | Writing and Communicative skills, Skills, Good Speaking ability, Knowledgeable, Highly analytical, Methodic, Creative, Commerce related Profession, Printing, Typing, Publishing, Mathematical Knowledge, Dry Land, Fond of Travel, Selected Food, History, Geography, Geology, Landed Property, Lawyer, Legal knowledge, Accountancy, Research, Pleasant Conversationalist, Acting, Cine Field, Photography, Show Business, Drama, Versatile |

## 7. Libra

| Angular Coverage | 181-210° |
|---|---|
| **Lordship** | Venus |
| **Exaltation** | Saturn |
| **Debilitation** | Sun |
| **Direction** | West |
| **Element** | Air |
| **Friends** | Saturn, Mercury, Rahu |
| **Enemies** | Jupiter, Mars, Moon |
| **Neutral** | Ketu |
| **Attributes** | Artistically inclined, Balanced, Enterprising, Pleasant, Friendly, Friendship, Flexible, Bold, Frank opinions, Expression of opinions without reservation, Lawyers, Judges, Eye for Beauty, Business, Cleanliness, Organized, Cine Field, Judicial, Learned, Chemicals, Prosperity, House, Vehicles, Pleasing Manners, Cine Field, Drama, Show Business, Treasury, Methodical |

## 8. Scorpio

| | |
|---|---|
| **Angular Coverage** | 211-240° |
| **Lordship** | Mars |
| **Exaltation** | None |
| **Debilitation** | Moon |
| **Direction** | North |
| **Element** | Water |
| **Friends** | Sun, Jupiter, Moon, Ketu |
| **Enemies** | Saturn, Venus, Mercury, Rahu |
| **Neutral** | None |
| **Attributes** | Machinery, Metals, Lands and Land Produce, Very Secretive, Profession related to secrecy, Watery, Wet Land, Pump set, Agricultural implements, Agriculture/Cultivation, Adventurous, Technical Skills, Instruments, Mining, Controlled Emotions, Sadness, Accountancy, Police, Armed Services, Engineer, Forceful, Machinery, Surgery, Medicine. |

## 9. Sagittarius

| | |
|---|---|
| **Angular Coverage** | 241-270° |
| **Lordship** | Jupiter |
| **Exaltation** | None |
| **Debilitation** | none |
| **Direction** | East |
| **Element** | Fire |
| **Friends** | Sun, Mars, Saturn, Ketu, |
| **Enemies** | Venus, Mercury, Moon, Rahu |
| **Neutral** | None |
| **Attributes** | Religious, Adherence to rituals, Conventional, Respect for Elders, Forest products, Priest, Teaching, Patience, Steady Prosperity, Turbulent and Struggling Mid-life, Later Period of life peaceful, Slow and Steady rise in Career, Success, Teacher (Guru), Spiritual, Academic, Research, Famous, Respected in society |

## 10. Capricorn

| | |
|---|---|
| **Angular Coverage** | 271-300° |
| **Lordship** | Saturn |
| **Exaltation** | Mars |
| **Debilitation** | Jupiter |
| **Direction** | South |
| **Element** | Land |
| **Friends** | Venus, Mercury, Rahu |
| **Enemies** | Sun, Moon, Mars |
| **Neutral** | Ketu |
| **Attributes** | Determined, Patience, Persistence, Perseverance, Mental Strength, Stubborn, Dull, People Contact, Politics, Dry Land, Landed Property, Mining, Metals, Judge, Research, Teaching, Travel, Hardware, Paints and Oils, Dependable, Reliable, Iron, Responsible, Truthful |

## 11. Aquarius

| | |
|---|---|
| **Angular Coverage** | 301-330° |
| **Lordship** | Saturn |
| **Exaltation** | None |
| **Debilitation** | None |
| **Direction** | West |
| **Element** | Air |
| **Friends** | Venus, Jupiter, Mercury, Rahu |
| **Enemies** | Mars, Sun, Moon |
| **Neutral** | Ketu |
| **Attributes** | Eccentric, Metaphysical, Mentally Evolved, Interest in Occult, Research, Intuitive, Broadcasting, Space, Air related, Drinks, Drugs, Medicine, Extra-ordinary level of understanding, Interested in hidden knowledge, Personality not fathomed easily, Delve into unknown |

## 12. Pisces

| | |
|---|---|
| **Angular Coverage** | 331-360° |
| **Lordship** | Jupiter |
| **Exaltation** | Venus |
| **Debilitation** | Mercury |
| **Direction** | North |
| **Element** | Water |
| **Friends** | Sun, Saturn, Mars, Moon, Ketu |
| **Enemies** | Venus, Mercury, Rahu |
| **Neutral** | None |
| **Attributes** | Highly Intuitive, General Prosperity assured, Metaphysical, Extra-ordinary level of understanding, Pleasure loving, Success with less effort, Research, Teaching, Spiritual life, Pump Sets, Agricultural activities, Religious. |

3.  It is now unequivocally established that characteristics of individuals are attributable to their genes. The visible attributes like appearance, speech, intellect and diseases, to name a few, are related to the genes. There is a correlation between planetary influence and the GMU of the individual. Planets represent attributes in the same way as Genetic materials – DeoxyriboNucleic Acid(DNA) and chromosomes do. Chromosomes are constituted by huge amount of genes made up of DNA molecules. Unless the GMU of the individual is in 'Resonance' with the Planet representing the same attributes, Genetic DNA/Chromosomes, cannot bring out the best in the individual and vice versa. That planetary combinations can also describe them is another clear indication, that planetary influences, play a decisive role in controlling genetic expressions in individuals. Expression of genes which result in bestowing certain characteristic and suppression of genes which even manifests at a later stage in life, also show that they could be regulated by planetary progressions. The genetic factor is the one, which bring out the differences

between individuals born at the same time, to different parents, at the same place, including the twins. The differences in the extent of expression should depend on the individual GMU which accounts for a difference of degree and not of the kind.

Here are examples of individuals who were born under the same Planetary configurations, bringing out similarities and Differences. The distinct differences are definitely due to variations in The GMU and the environment to which these two Individuals were subjected to:

## CHART 1

**Planets in Parenthese indicate Retrogression.**

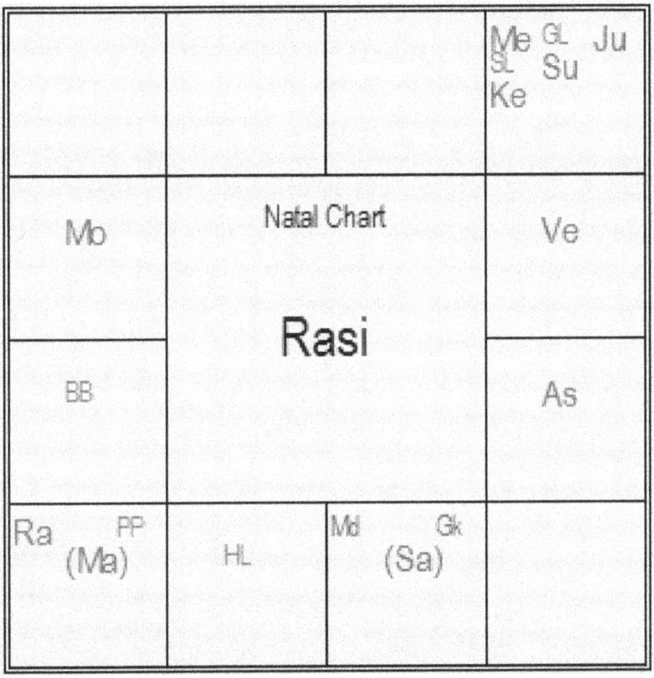

<table>
<tr><td></td><td></td><td></td><td>Me GL Ju<br>SL Su<br>Ke</td></tr>
<tr><td>Mo</td><td colspan="2" rowspan="2">Natal Chart<br><br>Rasi</td><td>Ve</td></tr>
<tr><td>BB</td><td>As</td></tr>
<tr><td>Ra      PP<br>(Ma)</td><td>HL</td><td>Md      Gk<br>(Sa)</td><td></td></tr>
</table>

|  | CASE 1 | CASE 2 |
| --- | --- | --- |
| **Native** | Intelligent. Possess creative skills Spiritual. | Intelligent. Possess creative skills. Spiritual. |
| **Father/Mother** | Father worked in a concern manufacturing films. | Father Employed in in an arts related profession. |
| **Brothers/Sisters** | Has an elder as well as younger sister. Younger sister suffered from Nervous dis-order. Both sisters led a moderate life. | Has sisters. Both sisters led a moderate life. |
| **Education** | Technical Education dealing with Machinery. | Good Education. Studied a para-medical subject dealing with micro-organisms. Later switched over to developing bio-materials based equipments for bio-chemical measurements. This happened when he visited a foreign country at 40-42. |
| **Marriage/Wife/ Husband** | Got married at 28-29. Has association with few woman. | Marriage around 28-29. |
| **Children** | A daughter and son. Both the Children came up well. | Has a daughter and son. One daughter took to paramedical line. A son did not do well. |
| **Occupation** | Dealt with machinery from 21 which produces automobile parts. Struggled initially. Started his own concern at 36-37. Did well. | Joined Government organisation as Researcher at 29-30. Collaborated with few industries. Got good name at work. Received few awards as well. |
| **Diseases** | More or less healthy. Suffere from nervous disorders after 54 | Suffered from blood pressure related problems from 42 onwards. Later from 54 suffered from nervous dis-orders. |
| **Finances/Affluence** | Reasonably prosperous. Possessed house, vehicles and land. | Reasonably prosperous. Possessed house, vehicles and land. |
| **Longevity** | Lived beyond 70 | Lived beyond 70 |

## CHART 2

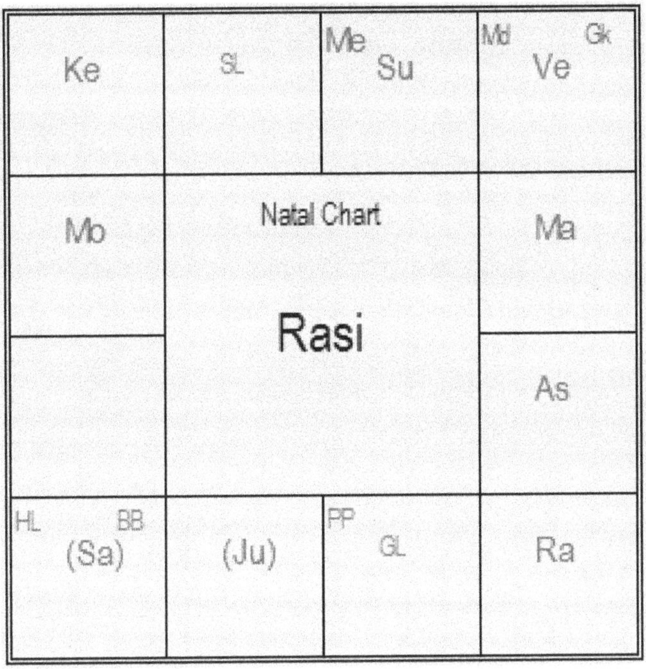

<table>
<tr><td>Ke</td><td>SL</td><td>Me  Su</td><td>Md  Gk<br>Ve</td></tr>
<tr><td>Mo</td><td colspan="2" rowspan="2">Natal Chart<br><br>Rasi</td><td>Ma</td></tr>
<tr><td></td><td>As</td></tr>
<tr><td>HL  BB<br>(Sa)</td><td>(Ju)</td><td>PP  GL</td><td>Ra</td></tr>
</table>

|  | CASE 1 | CASE 2 |
| --- | --- | --- |
| **Native** | Intelligent. | Intelligent. |
| **Father/Mother** | Father served as an official at a higher level in Government. Was also an agriculturist. Possessed lot of lands. | Father served at a lower level in a Government organization. Possessed few lands. |
| **Brothers/Sisters** | Only Child | Eldest. One elder and one younger sister and one younger brother. |
| **Education** | Technical Education dealing with Machinery. Higher education levels. Went abroad for higher education around 34 for a PhD degree. | Technical Education dealing with Machinery. Lower level of education. |

|  | CASE 1 | CASE 2 |
|---|---|---|
| **Marriage/Wife/ Husband** | Got married at 28. Has association with few woman. Wife left him and lived away from him for sometime. | Marriage around 28. Wife left him and lived away from him for sometime. |
| **Children** | A daughter and son. Both the Children came up well. | A daughter and son. Both did well in life. |
| **Occupation** | Dealt with Food machinery in a Government research organization. Came up with few inventions and fabrications. Published research papers. Patented equipments. Received few awards as well. | Dealt with Food machinery in a Government research organization. Came up with few inventions and fabrications. However did not get the deserved recognition. |
| **Diseases** | More or less healthy. Suffered diabetes and related problems from 46. | More or less healthy. Did not suffer much from BP and related problems. |
| **Finances/Affluence** | Highly prosperous. Inherited house, vehicles and plenty of land. | Reasonably prosperous. Possessed house, vehicles and land. |
| **Longevity** | Died at 58 due to kidney failure. | Lived beyond 70 |

A lot of similarity can be seen in these two cases. However, note the variations in these two cases as well.

This is the extent to which astrology can predict.

4.  It is also quite interesting to observe that the proposed effects for the planets in astrology (Sun, Moon, Mercury, Venus, Mars, Jupiter, and Saturn) are of the same magnitude - none too small, irrespective of the differences in masses and distances of these planets from the earth. Gravity, however, varies with mass and distance. All the planets, in astrology, exert the same extent of influence on the individual – none too small and none too large. Their huge difference in distance up to Saturn from Earth does not matter in astrology.

5.  In astrology, besides the planets, two distinct points in space (Moon's North Node and South Node - called Rahu and Ketu in Indian astrology), representing the two intersecting points in the orbital plane of the Earth to that of the Moon (orbiting around the Earth), are also considered important. They are treated as two planets. These points are used in calculations to compute the occurrence of Solar and Lunar Eclipses in astronomy, but they hold special significance in astrology.

For a student of astrology, the importance of these two points in predicting certain important aspects in the lives of people is well known. Rahu governs the ending of a period or a process, longevity, divergence, foreign residence, largeness of an operation, multitude, and distraction or obstruction in lives, among other things. Its counterpart, Ketu, signifies rope, bondage, convergence, narrowness, longevity, distraction or obstruction in lives, relief from mental, physical, legal, and professional troubles, ailments, tumor, cancer, and incarceration among other aspects. It can be recollected here that Clark Maxwell, while proposing his Theory of Electromagnetic radiation, described the Magnetic field as a Convergent one and the Electric field as a Divergent one. Rahu and Ketu represent one such Divergent and Convergent fields respectively.

These two intersecting points in space – Moon's North Node (Rahu) and South Node (Ketu) - play a pivotal role in functioning as gateways to the planetary influences. Starting from Distraction and Obstruction to CHAOS to Opportunities and to New Paths (Dimensions), they 'Wave Flag' (Ketu) or 'Terminate' (Rahu) a series of events in an individual's life. They, in fact, 'Run' the lives of individuals because our lives are nothing without these distractions and obstructions, good or bad.

Again, planets between the Rahu-Ketu axis and planets between the Ketu-Rahu axis do have different connotations in astrology!

Quite a lot of importance is attributed to the role of Rahu and Ketu in influencing the Life Propeller-Jupiter, and the Karma designator-Saturn, especially in the Naadi system of Indian Astrology. If astrology is not a science, why should two points in space, and that too, connected with the intersection points of the orbital planes of the Earth and Moon, assume so much significance, behaving much more than a planet?

## Diagrams Showing Rahu-Ketu Axis

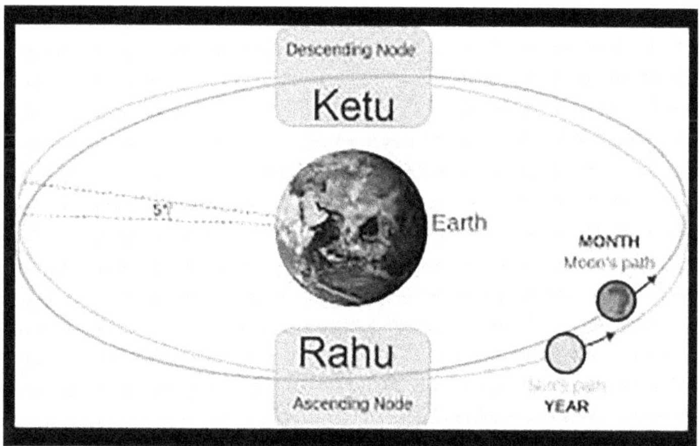


**Reproduced from www.karmafriendly.com**

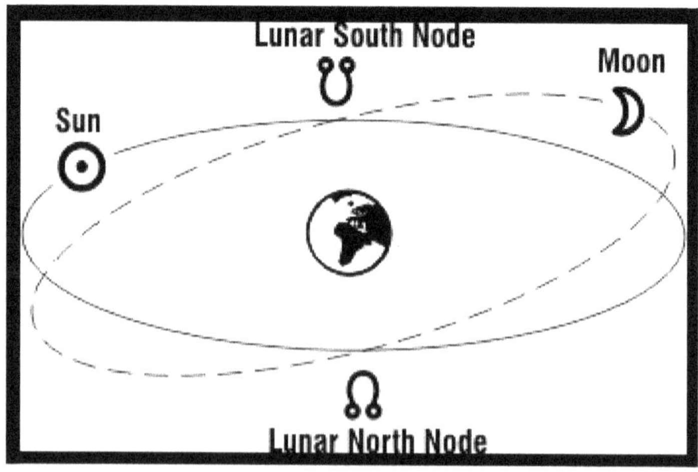

6.  Exchange of Planets: If a planet owning a sign is placed in another sign and if the lord of the sign is placed in the said planet's sign, then it is called a planetary exchange. Planetary exchanges play a crucial role in any horoscope. In many cases, such an exchange simplifies the prediction radically, as the exchange would describe the personality and the events related to the person distinctly with all the events dictated around that central dogma of exchange. The timing of the events from an exchange also holds the key to important events in a horoscope. Exchanges describe a change of place, change of profession, partners joining or deserting, desertion by a spouse or a friend, elopement, change in financial and professional status, birth of children, intervention or help from friend(s) or government, progress or suffering inlife from the time of exchange etc. These things could either happen to the individual or to the people associated with his life, like a wife, siblings, and parents, depending on the attributes of the associative planets.

# Exchanges involving Jupiter

**A.** **Jupiter-Saturn Exchange (only with respect to one house is shown) - Jupiter in Aquarius (Lord Saturn), Saturn in Pisces (Lord Jupiter)**

**B.** **Jupiter-Mars Exchange (only with respect to one house is shown) - Jupiter in Aries (Lord Mars), Mars in Sagittarius (Lord Jupiter)**

**Chart A**

<table>
<tr><td>Sat(11N)<br>Jup(19)</td><td>Jup(31)<br>Sat(43-72)</td><td>Jup(43)</td><td>Jup(55)</td></tr>
<tr><td>Jup(7N)<br>Sat(12-42)</td><td colspan="2" rowspan="2">A      Sat 11(N)<br><br>Jup 7(N)</td><td>Jup(67)</td></tr>
<tr><td></td><td></td></tr>
<tr><td></td><td></td><td></td><td></td></tr>
</table>

**Chart B**

<table>
<tr><td></td><td>Jup(2 N)<br>Mar(15-)</td><td>Jup(14)</td><td>Jup(38)</td></tr>
<tr><td></td><td colspan="2" rowspan="2">B      Jup2(N)</td><td></td></tr>
<tr><td></td><td></td></tr>
<tr><td>Mar(14N)<br>Jup(26)</td><td></td><td></td><td></td></tr>
</table>

**C.** **Chart 1 (Jupiter-Venus, Saturn-Moon, Sun-Mars Exchanges) - Jupiter in Taurus (Lord Venus), Venus in Sagittarius (Lord Jupiter) - Saturn in Cancer (Lord Moon), Moon in Capricorn (Lord Saturn) - Mars in Leo (Lord Sun), Sun in Scorpio (Lord Mars)**

<table>
<tr><td>Jup(54-65)</td><td>Jup(67)</td><td>Jup(R)(4.9)<br>Ven(18-67)</td><td>Ketu<br>Jup(5-17)</td></tr>
<tr><td>Sat(38-67)<br>Jup(42-53)</td><td colspan="2" rowspan="2">Chart 1<br><br><br><br>Female</td><td>Sat(N 7)<br>Moon(8-67)</td></tr>
<tr><td>Moon(7)<br>Sat(8-37)<br>Jup(30-41)</td><td>Mar(7)<br>Sun(8-67)</td></tr>
<tr><td>Jup(18-29)<br>Ven(N17)<br>Rahu</td><td>Sun(7)<br>Merc(67)<br>Mar(8-67)</td><td></td><td></td></tr>
</table>

**D. Jupiter-Mercury Exchange (only with respect to one house) - Jupiter in Gemini (Lord Mercury), Mercury in Sagittarius (Lord Jupiter)**

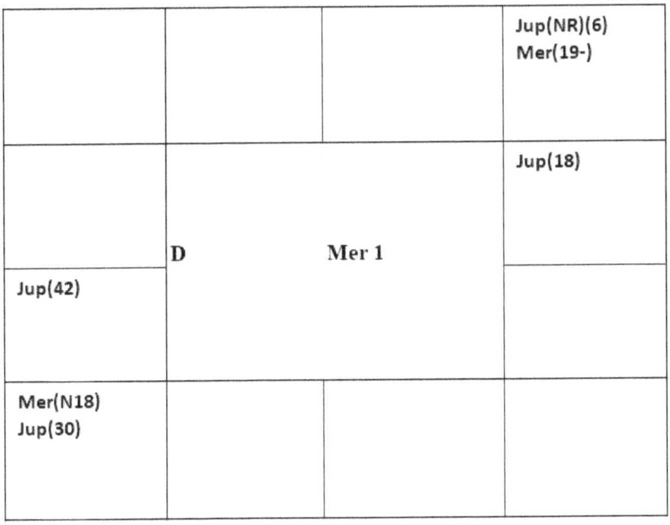

<table>
<tr><td></td><td></td><td></td><td>Jup(NR)(6)<br>Mer(19-)</td></tr>
<tr><td rowspan="2"></td><td rowspan="2" colspan="2">D      Mer 1</td><td>Jup(18)</td></tr>
<tr><td rowspan="2"></td></tr>
<tr><td>Jup(42)</td><td colspan="2"></td></tr>
<tr><td>Mer(N18)<br>Jup(30)</td><td></td><td></td><td></td></tr>
</table>

**E. Jupiter-Moon Exchange (only with respect to one house) - Jupiter in Cancer (Lord Moon), Moon in Pisces (Lord Jupiter)**

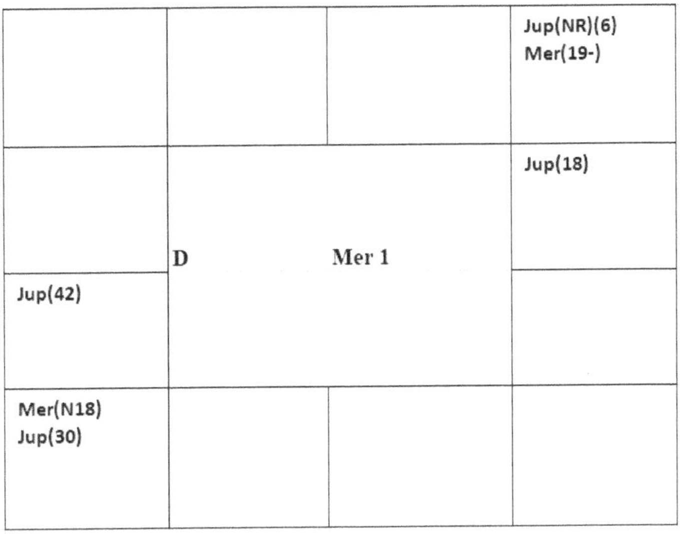

<table>
<tr><td></td><td></td><td></td><td>Jup(NR)(6)<br>Mer(19-)</td></tr>
<tr><td rowspan="2"></td><td rowspan="2" colspan="2">D      Mer 1</td><td>Jup(18)</td></tr>
<tr><td rowspan="2"></td></tr>
<tr><td>Jup(42)</td><td colspan="2"></td></tr>
<tr><td>Mer(N18)<br>Jup(30)</td><td></td><td></td><td></td></tr>
</table>

## F.  Jupiter-Sun Exchange (only with respect to one house) - Jupiter in Leo (Lord Sun), Sun in Sagittarius (Lord Jupiter)

| | | | |
|---|---|---|---|
| | | | |
| **F**<br>Jup(43) | | | Jup(N7)<br>Sun(20-) |
| Sun(N19)<br>Jup(31) | | | Jup(19) |

# FURTHER EXAMPLES (All the Exchanges)

## JUPITER - MERCURY EXCHANGES

**Mer 1** — **Jupiter in Gemini (Lord Mercury), Mercury in Pisces (Lord Jupiter)**

**Mercury 2** — **Jupiter in Gemini (Lord Mercury), Mercury in Sagittarius (Lord Jupiter)**

**Mercury 3** — **Jupiter in Virgo (Lord Mercury), Mercury in Sagittarius (Lord Jupiter)**

**Mercury 4** — **Jupiter in Virgo (Lord Mercury), Mercury in Pisces (Lord Jupiter)**

<table>
<tr><td>Mer(N 18)<br>Jup(30)<br>Moon<br>Ven</td><td></td><td></td><td>Jup(N 6)<br>Moon(6)<br>Ven(6)<br>Mer(19-)</td></tr>
<tr><td></td><td colspan="2" rowspan="2">Mer 1</td><td>Jup(18)<br>Moon<br>Ven</td></tr>
<tr><td></td><td>Jup(42)<br>Moon<br>Ven</td></tr>
<tr><td></td><td></td><td></td><td></td></tr>
</table>

<table>
<tr><td></td><td></td><td></td><td>Jup(N 7)<br>Moon(7)<br>Mer(20-)</td></tr>
<tr><td></td><td colspan="2" rowspan="2">Mer 2</td><td>Jup(19)<br>Moon</td></tr>
<tr><td>Jup(43)<br>Moon</td><td></td></tr>
<tr><td>Mer(N 19)<br>Jup(31)<br>Moon</td><td></td><td></td><td></td></tr>
</table>

<table>
<tr><td></td><td></td><td></td><td></td></tr>
<tr><td></td><td colspan="2" rowspan="2">Mer 3</td><td></td></tr>
<tr><td>Jup(42)<br>Moon<br>Ven</td><td></td></tr>
<tr><td>Mer(N 18)<br>Jup(30)<br>Moon<br>Ven</td><td></td><td>Jup(18)<br>Moon<br>Ven</td><td>Jup(N 6)<br>Moon(6)<br>Ven(6)<br>Mer(19-)</td></tr>
</table>

<table>
<tr><td>Mer(N 18)<br>Jup(30)<br>Moon<br>Ven</td><td>Jup(42)<br>Moon<br>Ven</td><td></td><td></td></tr>
<tr><td></td><td colspan="2" rowspan="2">Mer 4</td><td></td></tr>
<tr><td></td><td></td></tr>
<tr><td></td><td></td><td>Jup(18)<br>Moon<br>Ven</td><td>Jup(N 6)<br>Moon(6)<br>Ven(6)<br>Mer(19-)</td></tr>
</table>

# JUPITER- SATURN EXCHANGES

**Sat 1**  –  **Jupiter in Aquarius (Lord Saturn), Saturn in Pisces (Lord Jupiter)**

**Sat 2**  –  **Jupiter in Capricorn (Lord Saturn), Saturn in Pisces (Lord Jupiter)**

**Sat 3**  –  **Jupiter in Capricorn (Lord Saturn), Saturn in Sagittarius (Lord Jupiter)**

**Sat 4**  –  **Jupiter in Aquarius (Lord Saturn), Saturn in Sagittarius (Lord Jupiter)**

**Sat1**

| Sat(11N) Jup(19) | Jup(31) Sat(43-72) | Jup(43) | Jup(55) |
|---|---|---|---|
| Jup(7N) Sat(12-42) | Sat1     Sat 11(N) | | Jup(67) |
| | Jup 7(N) | | |
| | | | |

**Sat2**

| Sat(11N) Jup(31) | Jup(43) Sat(43-72) | | |
|---|---|---|---|
| Jup(19) | Sat2     Sat 11(N) | | |
| Jup(7N) Sat(12-42) | | | |
| | Jup 7(N) | | |

**Sat3**

| Jup(43) | Jup(55) | Jup(67) | Jup(79) |
|---|---|---|---|
| Jup(19) Sat(43-72) | Sat3     Sat 11(N) | | |
| Jup(N7) Sat(12-42) | | | |
| Sat(11) Jup(31) | Jup 7(N) | | |

**Sat4**

| Jup(19) | Jup(55) | Jup(67) | |
|---|---|---|---|
| Jup(N7) Sat(12-42) | Sat4     Sat 11(N) | | |
| Sat(43-72) Jup(43) | | | |
| Sat(11) Jup(31) | Jup 7(N) | | |

# JUPITER- MARS EXCHANGES

**Mars 1**  –  **Jupiter in Aries (Lord Mars), Mars in Pisces (Lord Jupiter)**

**Mars 2**  –  **Jupiter in Aries (Lord Mars), Mars in Sagittarius (Lord Jupiter)**

**Mars 3**  –  **Jupiter in Scorpio (Lord Mars), Mars in Sagittarius (Lord Jupiter)**

**Mars 4**  –  **Jupiter in Scorpio (Lord Mars), Mars in Pisces (Lord Jupiter)**

<table>
<tr><td>Mar(14N)<br>Jup(26)</td><td>Jup(2 N)<br>Mar(15-)</td><td>Jup(14)</td><td>Jup(38)</td></tr>
<tr><td></td><td colspan="2" rowspan="2">Mars 1     Jup2(N)</td><td>Jup(50)</td></tr>
<tr><td></td><td></td></tr>
<tr><td></td><td></td><td></td><td></td></tr>
</table>

<table>
<tr><td></td><td>Jup(2 N)<br>Mar(15-)</td><td>Jup(14)</td><td>Jup(38)</td></tr>
<tr><td></td><td colspan="2" rowspan="2">Mars 2     Jup2(N)</td><td></td></tr>
<tr><td></td><td></td></tr>
<tr><td>Mar(14N)<br>Jup(26)</td><td></td><td></td><td></td></tr>
</table>

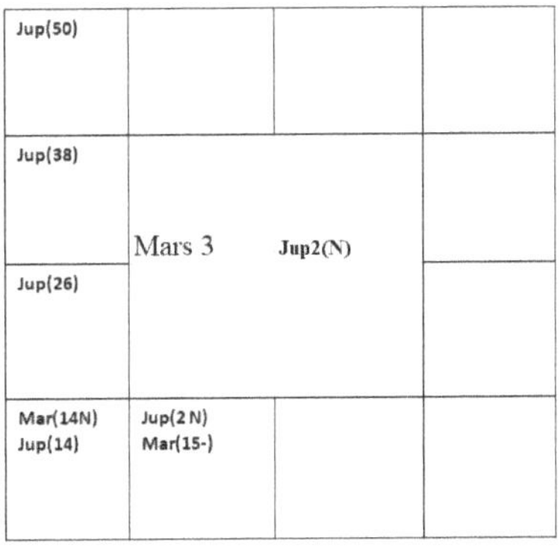

<table>
<tr><td>Jup(50)</td><td></td><td></td><td></td></tr>
<tr><td>Jup(38)</td><td colspan="2" rowspan="2">Mars 3     Jup2(N)</td><td></td></tr>
<tr><td>Jup(26)</td><td></td></tr>
<tr><td>Mar(14N)<br>Jup(14)</td><td>Jup(2 N)<br>Mar(15-)</td><td></td><td></td></tr>
</table>

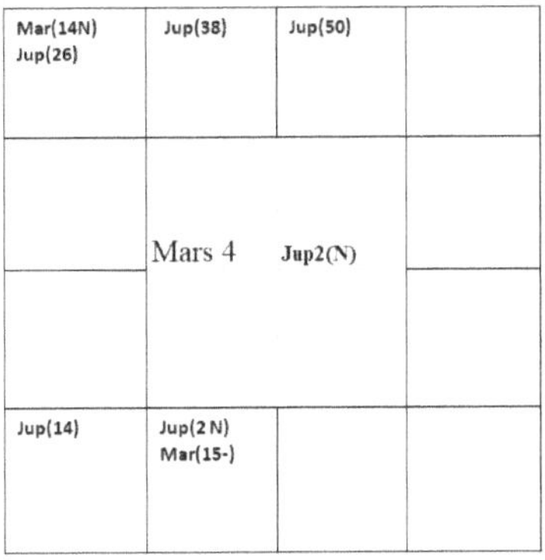

<table>
<tr><td>Mar(14N)<br>Jup(26)</td><td>Jup(38)</td><td>Jup(50)</td><td></td></tr>
<tr><td></td><td colspan="2" rowspan="2">Mars 4     Jup2(N)</td><td></td></tr>
<tr><td></td><td></td></tr>
<tr><td>Jup(14)</td><td>Jup(2 N)<br>Mar(15-)</td><td></td><td></td></tr>
</table>

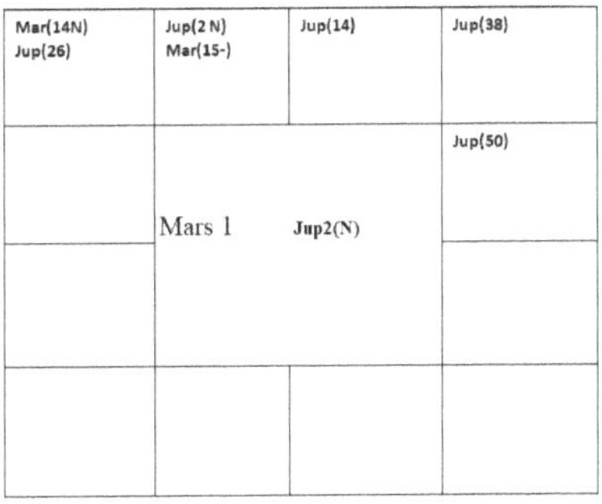

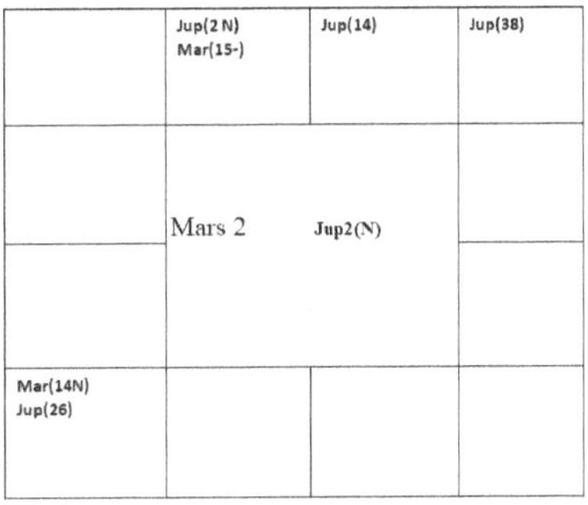

# JUPITER- VENUS EXCHANGES

**Ven 1  –  Jupiter in Taurus (Lord Venus), Venus in Pisces (Lord Jupiter)**

**Ven 2  –  Jupiter in Taurus (Lord Venus), Venus in Sagittarius (Lord Jupiter)**

**Ven 3  –  Jupiter in Libra (Lord Venus), Venus in Pisces (Lord Jupiter)**

**Ven 4  –  Jupiter in Libra (Lord Jupiter), Venus in Sagittarius (Lord Jupiter)**

**Ven 1**

| | | | |
|---|---|---|---|
| Ven(N20) Jup(32) | | Jup(N8) Ven(21-) | Jup(20) |
| | **Ven 1** | | Jup(44) |
| | | | |
| | | | |

**Ven 2**

| | | | |
|---|---|---|---|
| | | Jup(N 8) Ven(21-) | Jup(20) |
| Jup(44) | **Ven 2** | | |
| Ven(N20) Jup(32) | | | |

**Ven 3**

| | | | |
|---|---|---|---|
| Ven(N 20) Jup(32) | Jup(44) | | |
| | **Ven 3** | | |
| | Jup(20) | Jup(N8) Ven(21-) | |

**Ven 4**

| | | | |
|---|---|---|---|
| Jup(68) | | | |
| Jup(56) | **Ven 4** | | |
| Jup(44) | | | |
| Ven(N 20) Jup(32) | Jup(20) | Jup(N8) Ven(21-) | |

## JUPITER- SUN EXCHANGES

**Sun 1   –   Jupiter in Leo (Lord Sun), Sun in Sagittarius (Lord Jupiter)**

**Sun 2   –   Jupiter in Leo (Lord Sun), Sun in Pisces (Lord Jupiter)**

| | | | |
|---|---|---|---|
| | | | |
| Sun 1 | | | Jup(N7)<br>Sun(20-) |
| Jup(43) | | | |
| Sun(N19)<br>Jup(31) | | | Jup(19) |

| | | | |
|---|---|---|---|
| Sun(N19)<br>Jup(31) | Jup(43) | | |
| | Sun 2 | | Jup(N7)<br>Sun(20-) |
| | | | Jup(19) |

## JUPITER- MOON EXCHANGES

**Moon 1 – Jupiter in Cancer (Lord Moon), Moon in Pisces (Lord Jupiter)**

**Moon 2 – Jupiter in Cancer (Lord Moon), Moon in Sagittarius (Lord Jupiter)**

| | | | |
|---|---|---|---|
| Moon(N18)<br>Jup(30) | | | |
| | Moon 1    Jup6(N) | | Moon(19-)<br>Jup(N 6) |
| | | | Jup(18) |
| | | | Jup(42) |

| | | | |
|---|---|---|---|
| | | | |
| | Moon 2    Jup6(N) | | Moon(19-)<br>Jup(N 6) |
| Jup(42) | | | Jup(18) |
| Moon(N18)<br>Jup(30) | | | |

# Exchanges involving Saturn:

**A.** **Saturn-Sun Exchange - Saturn in Leo (Lord Sun), Sun in Capricorn (Lord Saturn)**

**B.** **Saturn-Mars Exchange - Saturn in Aries (Lord Mars), Mars in Aquarius (Lord Saturn)**

Chart A:

| | | | |
|---|---|---|---|
| | | | |
| Sat(55-84) | | Sat(24) Leo | |
| | A | | |
| Sun(N 24) Sat(25-54) | | Sat(N 24) Sun(25-) | |
| | | | |

Chart B:

| | | | |
|---|---|---|---|
| | Sat(11N) Mar(12-41) | Sat(42-71) | |
| Mar(11N) Sat(12-41) | | | |
| | B | Sat(N)11 | |
| | | | |
| | | | |

**C.** **Saturn-Venus Exchange - Saturn in Taurus (Lord Venus), Venus in Aquarius (Lord Saturn)**

**D.** **Saturn-Mercury Exchange - Saturn in Virgo (Lord Mercury), Mercury in Capricorn (Lord Saturn)**

Chart C:

| | | | |
|---|---|---|---|
| | | Sat( Natal, 0-26) Ven(27-) | Sat(57-87) |
| Ven( Natal 0-26) Sat(27-56 ) | | | |
| | C | | |
| | | | |
| | | | |

Chart D:

| | | | |
|---|---|---|---|
| | | | |
| Sat(56-85) | | | |
| | D | Sat(N)25 | |
| Mer(N 25) Sat(26-55) | | | |
| | | | Sat(N)(25) Mer(26-) |

## FURTHER EXAMPLES (All the Exchages)

### SATURN – MARS EXCHANGES

**Mars 1  –  Saturn in Aries (Lord Mars), Mars in Aquarius (Lord Mars) Saturn)**

**Mars 2  –  Saturn in Aries (Lord Mars), Mars in Capricorn (Lord Mars) Saturn)**

**Mars 3  –  Saturn in Scorpio (Lord Mars), Mars in Capricorn (Lord Saturn)**

**Mars 4  –  Saturn in Scorpio (Lord Mars), Mars in Aquarius (Lord Saturn)**

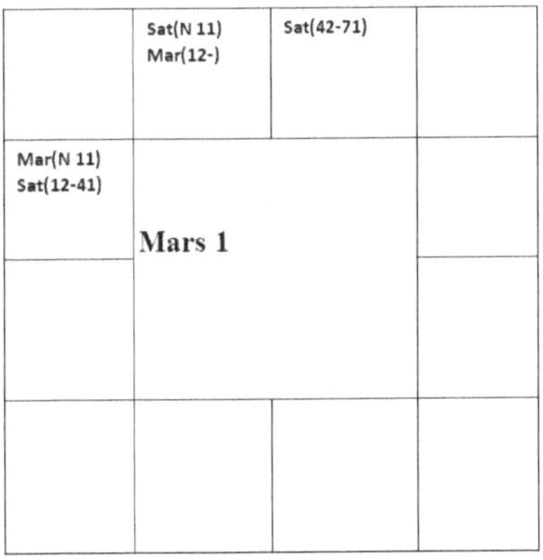

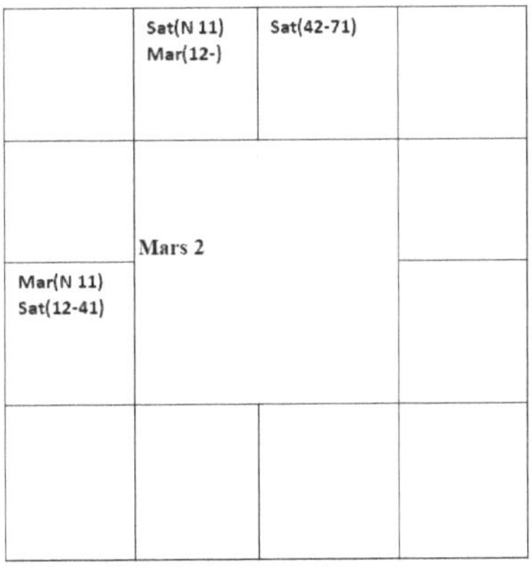

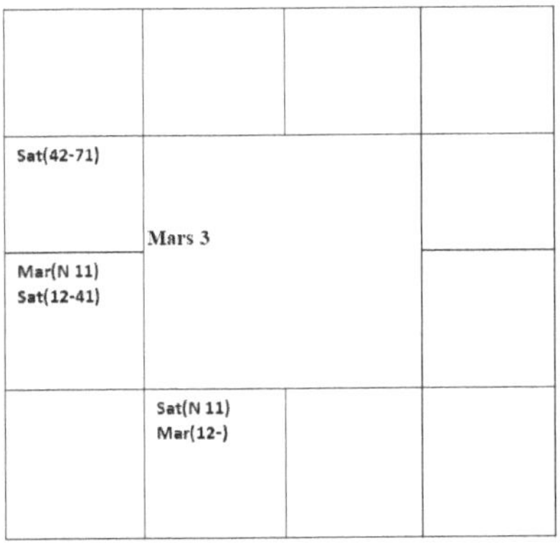

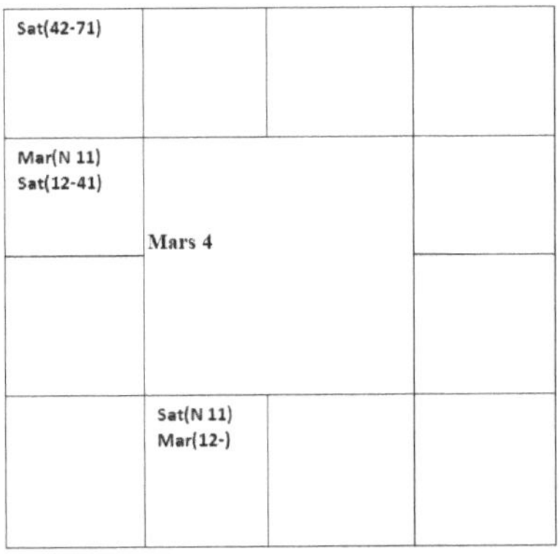

**Mars 1**

Sat(N 11) Mar(12-) ; Sat(42-71) ; Mar(N 11) Sat(12-41)

**Mars 2**

Sat(N 11) Mar(12-) ; Sat(42-71) ; Mar(N 11) Sat(12-41)

**Mars 3**

Sat(42-71) ; Mar(N 11) Sat(12-41) ; Sat(N 11) Mar(12-)

**Mars 4**

Sat(42-71) ; Mar(N 11) Sat(12-41) ; Sat(N 11) Mar(12-)

# SATURN – VENUS EXCHANGES

**Venus 1 – Saturn in Taurus (Lord Venus), Venus in Aquarius (Lord Saturn)**

**Venus 2 – Saturn in Taurus (Lord Venus), Venus in Capricorn (Lord Saturn)**

**Venus 3 – Saturn in Libra (Lord Venus), Venus in Capricorn (Lord Saturn)**

**Venus 4 – Saturn in Libra (Lord Venus), Venus in Aquarius (Lord Saturn)**

**Venus 1**

| | | Sat( N 26) Ven(27-) | Sat(57-87) |
|---|---|---|---|
| Ven( N 26) Sat(27-56 ) | | | |
| | Venus 1 | | |
| | | | |

**Venus 2**

| | | Sat( N 26) Ven(27-) | Sat(57-87) |
|---|---|---|---|
| | | | |
| Ven( N 26) Sat(27-56 ) | Venus 2 | | |
| | | | |

**Venus 3**

| | | | |
|---|---|---|---|
| Sat( 57-87) | | | |
| Ven( N 26) Sat(27-56 ) | Venus 3 | | |
| | | Sat( N 26) Ven(27-) | |

**Venus 4**

| Sat( 57-87) | | | |
|---|---|---|---|
| Ven( N 26) Sat(27-56 ) | | | |
| | Venus 4 | | |
| | | Sat( N 26) Ven(27-) | |

# SATURN – MERCURY EXCHANGES

**Mer 1** – Saturn in Virgo (Lord Mercury), Mercury in Capricorn (Lord Saturn)

**Mer 2** – Saturn in Virgo (Lord Mercury), Mercury in Aquarius (Lord Saturn)

**Mer 3** – Saturn in Gemin (Lord Mercury), Mercury in Aquarius (Lord Saturn)

**Mer 4** – Saturn in Gemini (Lord Mercury), Mercury in Capricorn (Lord Saturn)

**Mer 1**

| | | | |
|---|---|---|---|
| | | | |
| Sat (56-85) | | | |
| Mer(N 25) Sat(26-55) | | | Sat (N 25) Mer(26-55) Mer(56-) |
| | | | |

**Mer 2**

| | | | |
|---|---|---|---|
| Sat (56-85) | | | |
| Mer(N 26) Sat(26-55) | | | |
| | | | Sat (N 25) Mer(26-55 Mer(56-) |
| | | | |

**Mer 3**

| | | | |
|---|---|---|---|
| | | | Sat (N 25) Mer(26-55) Mer(56-) |
| Mer(N 25) Sat(26-55) | | | Sat(56-85) |
| | | | |
| | | | |

**Mer 4**

| | | | |
|---|---|---|---|
| | | | Sat (N 25) Mer(26-55) Mer(56-) |
| | | | Sat(56-85) |
| Mer(N 25) Sat(26-55) | | | |
| | | | |

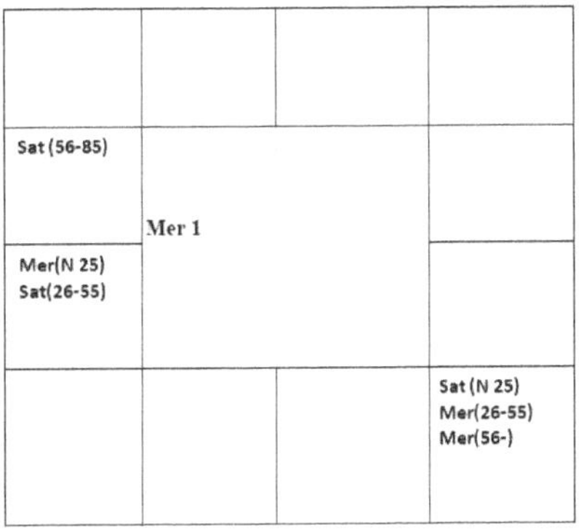

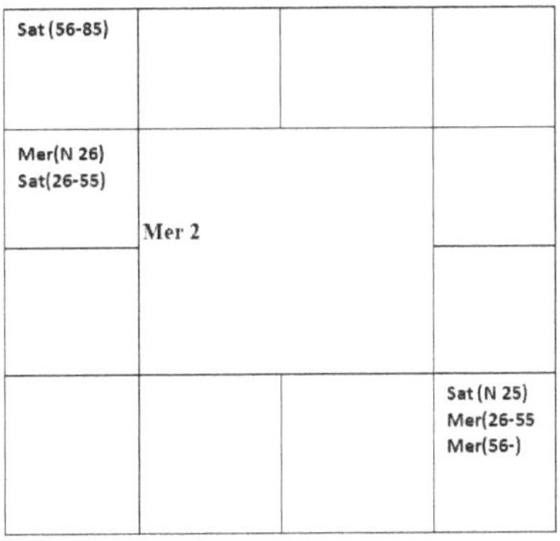

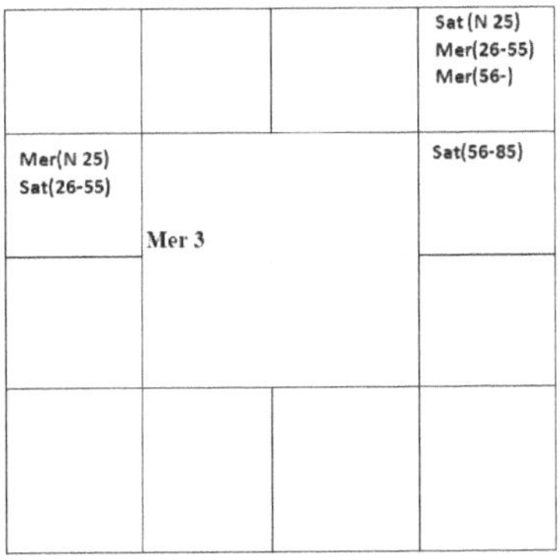

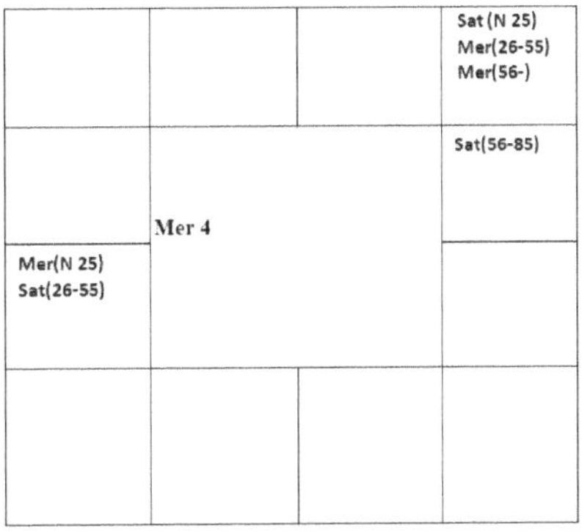

## SATURN – SUN EXCHANGES

**Sun 1  –  Saturn in Leo (Lord Sun), Sun in Capricorn (Lord Saturn)**

**Sun 2  –  Saturn in Leo (Lord Sun), Sun in Aquarius (Lord Saturn)**

<table>
<tr><td></td><td></td><td></td><td></td></tr>
<tr><td>Sat(55-84)<br><br>Sun (N 24)<br>Sat(25-54)</td><td colspan="2" rowspan="2">Sun 1</td><td></td></tr>
<tr><td></td><td>Sat(N 24)<br>Sun(25-)</td></tr>
<tr><td></td><td></td><td></td><td></td></tr>
</table>

<table>
<tr><td>Sat(55-84)</td><td></td><td></td><td></td></tr>
<tr><td>Sun (N 24)<br>Sat(25-54)</td><td colspan="2" rowspan="2">Sun 2</td><td></td></tr>
<tr><td></td><td>Sat(N 24)<br>Sun(25-)</td></tr>
<tr><td></td><td></td><td></td><td></td></tr>
</table>

## SATURN – MOON EXCHANGES

**Moon 1  –  Saturn in Cancer (Lord Moon), Moon in Capricorn (Lord Saturn)**

**Moon 2  –  Saturn in Cancer (Lord Moon), Moon in Aquarius (Lord Saturn)**

<table>
<tr><td></td><td></td><td></td><td></td></tr>
<tr><td>Sat(38-67)<br><br>Moon(N 7)<br>Sat(8-37)</td><td colspan="2" rowspan="2">Moon 1</td><td>Sat(N 7)<br>Moon (8-)</td></tr>
<tr><td></td></tr>
<tr><td></td><td></td><td></td><td></td></tr>
</table>

<table>
<tr><td>Sat(38-67)</td><td></td><td></td><td></td></tr>
<tr><td>Moon(N 7)<br>Sat(8-37)</td><td colspan="2" rowspan="2">Moon 2</td><td>Sat(N 7)<br>Moon (8-)</td></tr>
<tr><td></td></tr>
<tr><td></td><td></td><td></td><td></td></tr>
</table>

# Exchanges involving Mars:

**A.  Mars-Moon Exchange:**

**M-Moon 1 - Mars in Cancer (Lord Moon), Mars in Aries  (Lord Mars)**

**M-Moon 2 - Mars  in  Cancer  (Lord  Moon),  Mars  in  Scorpio  (Lord Mars)**

<table>
<tr><td colspan="3">

Moon(N 25)<br>Mar(26-)

</td></tr>
<tr><td></td><td>M-Moon 1    SAT 25 yrs</td><td>Mar(N 25)<br>Moon(26-)</td></tr>
<tr><td></td><td></td><td></td></tr>
</table>

<table>
<tr><td></td><td></td><td>Mar(N 25)<br>Moon(26-)</td></tr>
<tr><td>M-Moon 2    SAT 25 yrs</td><td></td><td></td></tr>
<tr><td>Moon(N 25)<br>Mar(26-)</td><td></td><td></td></tr>
</table>

**B.  Mars-Venus Exchange:**

**M-V1 – Mars in Libra (Lord Venus), Venus in Aries (Lord Mars)**

**M-V2 – Mars in Taurus (Lord Venus), Venus in Scorpio (Lord Mars)**

**M-V3 – Mars in Taurus (Lord Venus), Venus in Aries (Lord Mars)**

**M-V4 – Mars in Libra (Lord Venus), Venus in Scorpio (Lord Mars)**

<table>
<tr><td colspan="3">

Ven(N 21)<br>Mar(22-)

</td></tr>
<tr><td>M-V1    SAT 21yrs</td><td></td><td></td></tr>
<tr><td></td><td>Mar(N 21)<br>Ven(22-)</td><td></td></tr>
</table>

<table>
<tr><td></td><td>Mar(N 21)<br>Ven(22-)</td><td></td></tr>
<tr><td>M-V2    SAT 21yrs</td><td></td><td></td></tr>
<tr><td></td><td>Ven(N 21)<br>Mar(22-)</td><td></td></tr>
</table>

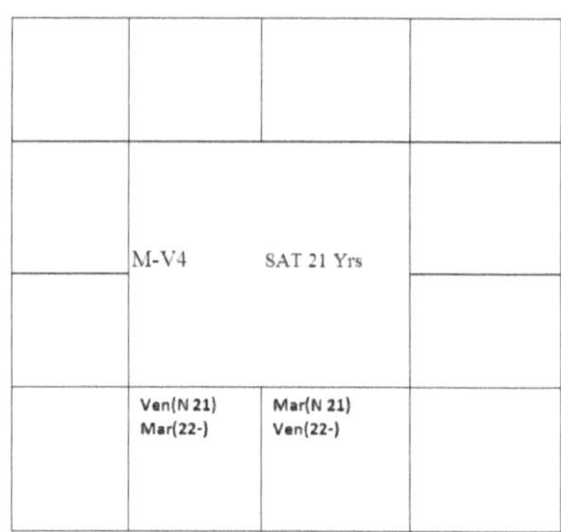

M-V3 — SAT 21 yrs: Ven(N 21) Mar(22-), Mar(N 21) Ven(22-)

M-V4 — SAT 21 Yrs: Ven(N 21) Mar(22-), Mar(N 21) Ven(22-)

## C.  Mars-Mercury Exchange:

**M-Mer 1 – Mars in Gemini (Lord Mercury), Mercury in Scorpio (Lord Mars)**

**M-Mer 2 – Mars in Virgo (Lord Mercury), Mercury in Scorpio (Lord Mars)**

**M-Mer 3 – Mars in Virgo (Lord Mercury), Mercury in Aries (Lord Mars)**

**M-Mer 4 – Mars in Gemini (Lord Mercury), Mercury in Aries (Lord Mars)**

M-Mer 1 — SAT 17 Yrs: Mar(N 17) Mer(18-), Mer(N 17) Mar18-)

M-Mer2 — SAT 17 Yrs: Mar(N 17) Mer(18-), Mar(N 17) Mer(18-)

<table>
<tr><td></td><td>Mer(N 17)<br>Mar(18-)</td><td></td><td></td></tr>
<tr><td rowspan="2">M-Mer3        SAT 17 Yrs</td><td colspan="2" rowspan="2"></td><td></td></tr>
<tr><td></td></tr>
<tr><td></td><td></td><td>Mar(N 17)<br>Mer(18-)</td></tr>
</table>

<table>
<tr><td></td><td>Mer(N 17)<br>Mar(18-)</td><td></td><td>Mar(N 17)<br>Mer(18-)</td></tr>
<tr><td rowspan="2">M-Mer4        SAT 17 Yrs</td><td colspan="2" rowspan="2"></td><td></td></tr>
<tr><td></td></tr>
<tr><td></td><td></td><td></td></tr>
</table>

## D.  Mars-Sun Exchange:

### M-Sun 1 – Mars in Leo (Lord Sun), Suns in Leo (Lord Mars)

### M-Sun 2 – Mars in Leo (Lord Sun), Suns in Scorpio (Lord Mars)

<table>
<tr><td></td><td>Sun(N 19)<br>Mars(20-)</td><td></td><td></td></tr>
<tr><td rowspan="2">M-Sun 1        SAT 19 Yrs</td><td colspan="2" rowspan="2"></td><td></td></tr>
<tr><td>Mar(N 19)<br>Sun (20-)</td></tr>
<tr><td></td><td></td><td></td></tr>
</table>

<table>
<tr><td></td><td></td><td></td><td></td></tr>
<tr><td rowspan="2">M-Sun2        SAT 19 Yrs</td><td colspan="2" rowspan="2"></td><td></td></tr>
<tr><td>Mar(N 19)<br>Sun (20-)</td></tr>
<tr><td>Sun(N 19)<br>Mars(20-)</td><td></td><td></td></tr>
</table>

# Exchanges involving Venus:

## A. Venus-Moon Exchange:

**V-M1 – Venus in Cancer (Lord Moon), Moon in Libra (Lord Venus)**

**V-M2 – Venus in Cancer (Lord Moon), Moon Taurus (Lord Venus)**

<table>
<tr><td></td><td></td><td></td><td></td></tr>
<tr><td></td><td rowspan="2">V-M1   SAT 20 Yrs</td><td>Ven(N 20)<br>Moon(21-)</td></tr>
<tr><td></td><td></td></tr>
<tr><td></td><td>Moon(N 20)<br>Ven(21-)</td><td></td></tr>
</table>

<table>
<tr><td></td><td>Moon(N 20)<br>Ven(21-)</td><td></td></tr>
<tr><td></td><td rowspan="2">V-M2   SAT 20 Yrs</td><td>Ven(N 20)<br>Moon(21-)</td></tr>
<tr><td></td><td></td></tr>
<tr><td></td><td></td><td></td></tr>
</table>

## B. Venus-Sun Exchange:

**V-S1 – Venus in Leo (Lord Sun), Sun in Libra (Lord Venus)**

**V-S2 – (Rare combination)– Venus in Leo (Lord Sun), Sun Taurus (Lord Venus)**

<table>
<tr><td></td><td></td><td></td></tr>
<tr><td></td><td rowspan="2">V-S1   SAT 16 Yrs</td><td></td></tr>
<tr><td></td><td>Ven(N 16)<br>Sun(17-)</td></tr>
<tr><td></td><td>Sun(N 16)<br>Ven(17-)</td><td></td></tr>
</table>

<table>
<tr><td></td><td>Sun(N 16)<br>Ven(17-)</td><td></td></tr>
<tr><td></td><td rowspan="2">V-S2   SAT 16 Yrs</td><td></td></tr>
<tr><td></td><td>Ven(N 16)<br>Sun(17-)</td></tr>
<tr><td></td><td></td><td></td></tr>
</table>

**C. Venus-Mercury Exchange:**

**V-ME1 – Venus in Virgo (Lord Mercury), Mercury in Libra (Lord Venus)**

**V-ME2 – Venus in Gemini (Lord Mercury), Mercury in Taurus (Lord Venus)**

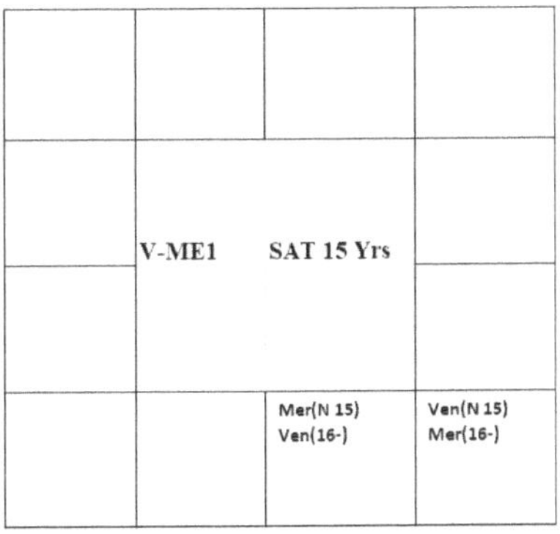

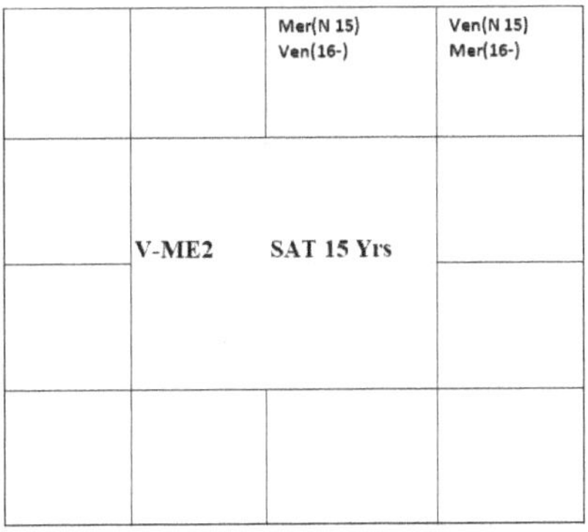

| | | | |
|---|---|---|---|
| | | | |
| | V-ME1    SAT 15 Yrs | | |
| | | Mer(N 15)<br>Ven(16-) | Ven(N 15)<br>Mer(16-) |

| | | | |
|---|---|---|---|
| | | Mer(N 15)<br>Ven(16-) | Ven(N 15)<br>Mer(16-) |
| | V-ME2    SAT 15 Yrs | | |
| | | | |

# Exchanges involving Mercury:

**A. Mercury-Moon Exchange:**

**ME-MO1 – Mercury in Cancer (Lord Moon), Moon in Virgo (Lord Mercury)**

**ME-MO2 – Mercury in Cancer (Lord Moon), Moon in Gemini (Lord Mercury)**

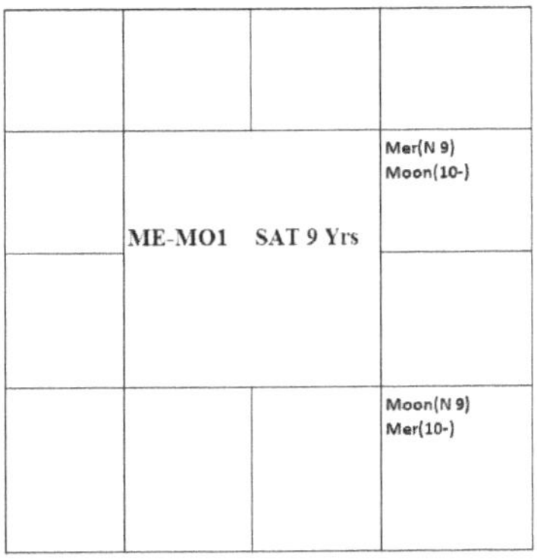

| | | | |
|---|---|---|---|
| | | | |
| | | | Mer(N 9)<br>Moon(10-) |
| | ME-MO1    SAT 9 Yrs | | |
| | | | Moon(N 9)<br>Mer(10-) |

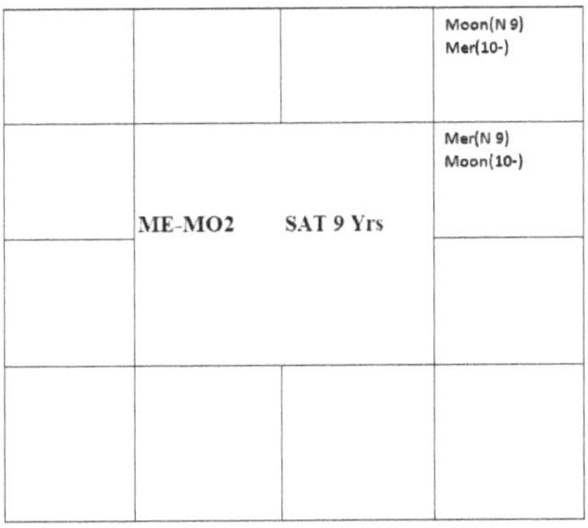

| | | | |
|---|---|---|---|
| | | | Moon(N 9)<br>Mer(10-) |
| | | | Mer(N 9)<br>Moon(10-) |
| | ME-MO2    SAT 9 Yrs | | |
| | | | |

## B.  Mercury-Sun Exchange:

### S-ME1 – Mercury in Leo (Lord Sun), Sun in Virgo (Lord Mercury)

### S-ME2– Mercury in Leo (Lord Sun), Sun in Gemini (Lord Mercury)

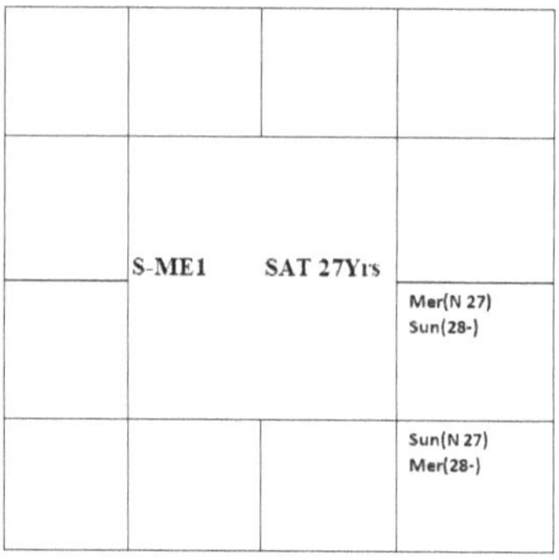

<table>
<tr><td></td><td></td><td></td><td></td></tr>
<tr><td></td><td colspan="2">S-ME1      SAT 27Yrs</td><td></td></tr>
<tr><td></td><td></td><td>Mer(N 27)<br>Sun(28-)</td><td></td></tr>
<tr><td></td><td></td><td>Sun(N 27)<br>Mer(28-)</td><td></td></tr>
</table>

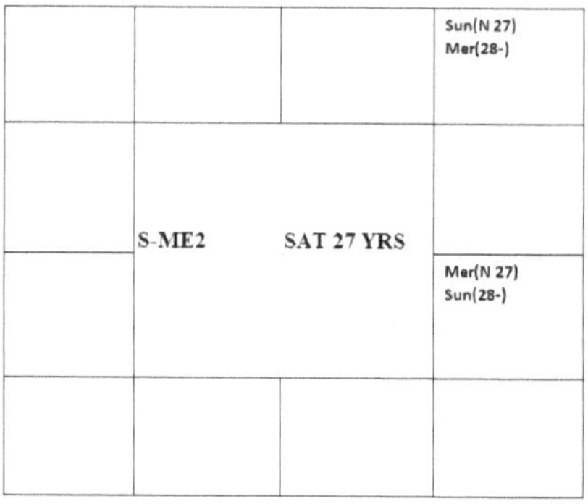

<table>
<tr><td></td><td></td><td></td><td>Sun(N 27)<br>Mer(28-)</td></tr>
<tr><td></td><td colspan="2">S-ME2      SAT 27 YRS</td><td></td></tr>
<tr><td></td><td></td><td>Mer(N 27)<br>Sun(28-)</td><td></td></tr>
<tr><td></td><td></td><td></td><td></td></tr>
</table>

## Exchange involving Sun and Moon:

## Moon in Leo (Lord Sun), Sun in Cancer (Lord Moon)

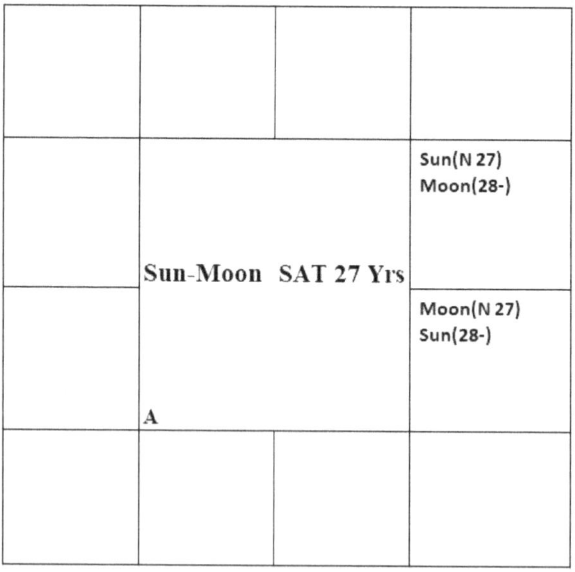

<table>
<tr><td></td><td></td><td></td><td></td></tr>
<tr><td></td><td rowspan="2">Sun-Moon  SAT 27 Yrs</td><td></td><td>Sun(N 27)<br>Moon(28-)</td></tr>
<tr><td></td><td></td><td>Moon(N 27)<br>Sun(28-)</td></tr>
<tr><td></td><td>A</td><td></td><td></td></tr>
<tr><td></td><td></td><td></td><td></td></tr>
</table>

The examples we saw above deal with regular exchanges.

There are also other types of Exchanges.

## CYCLIC EXCHANGES:

**R stands for retrogression.**

**Chart 1: Jupiter in the house of Mercury; Mercury-Saturn;**

**Saturn-Venus; Venus-Jupiter LINK in a CYCLIC manner made him a great Technocrat Manager.**

| Moon | Mar | | Jup(1)<br>Ketu |
|---|---|---|---|
| Sun | | | |
| Mer(R) | First Year | | |
| Ven<br>Rahu | | Sat(1)<br>Male | |

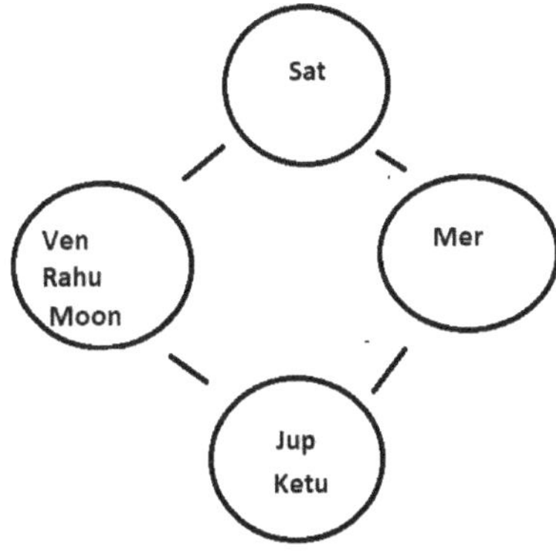

**Chart 2:** Jupiter in Cancer, Moon in Libra, Venus in Saturn, Saturn in Jupiter. The native is a great music composer who received accolades and awards for his work.

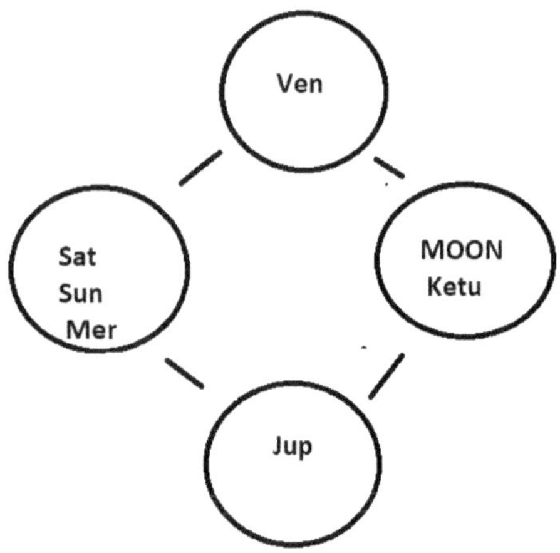


| Sat(28) | Rahu | | |
|---|---|---|---|
| | | | Jup(R 8) |
| Ven | | | |
| Sun<br>Mer | | Ketu<br>Moon<br>Jup(24) | Mar |

## PROGRESSIVE EXCHANGES:

These exchanges occur when Jupiter/Saturn are in progression.

**Chart 1**: Two Progressional Exchange involving Cancer-Sagittariusand later Capricorn-Pisces makes Jupiter complete the entire Zodiac in 84 years beginning from 17 years, illustrating the enormous success he achieved through his creative abilities.

| | | | |
|---|---|---|---|
| | Ketu | Jup(5) | Jup(17) |
| Sun<br>Ven<br>Sat(18) | 0-18 | | Jup(29) |
| Mer | | Male | |
| Mar<br>Moon | | Rahu | |

| | | | |
|---|---|---|---|
| Sun(48)<br>Ven(48)<br>Sat(48)<br>Jup(65) | Ketu<br>Jup(77)<br>Sun(79-)<br>Ven(79-)<br>Sat(79-)<br>Mar(79-) | | |
| Jup(53)<br>Mer(84)<br>Sun(49-78)<br>Ven(49-78)<br>Sat(49-78) | 49-84 | | Jup(29)<br>Moon(30-) |
| Jup(41)<br>Mar(77 )<br>Moon(29)<br>Jup(84) | | Rahu | |

**Chart 2:** This is an instance where Progressional Exchange between Jupiter and Mercury resulted in change of country and enhanced prosperity.

| | | | |
|---|---|---|---|
| | | | Ketu |
| | Natal    0-20 | | |
| Sat(20) | | | Jup(4) |
| | Sat 20 | Male | |
| Mer(R)<br>Rahu<br>Jup(28)<br>Moon(28) | Sun<br>Mar | Ven | Jup(16)<br>Moon<br>Mer(17-28) |

**Chart 3**: Exchange between Sun and Venus and further Progressional Cyclic exchange between Jupiter-Scorpio, Mars-Cancer and Moon-Pisces ensures enormous Political success to this Native.

| | | | |
|---|---|---|---|
| Moon | | Jup(R) | Rahu |
| Sat(R) | | | Mar |
| | | | Ven |
| Ketu | | Sun<br>Mer | |

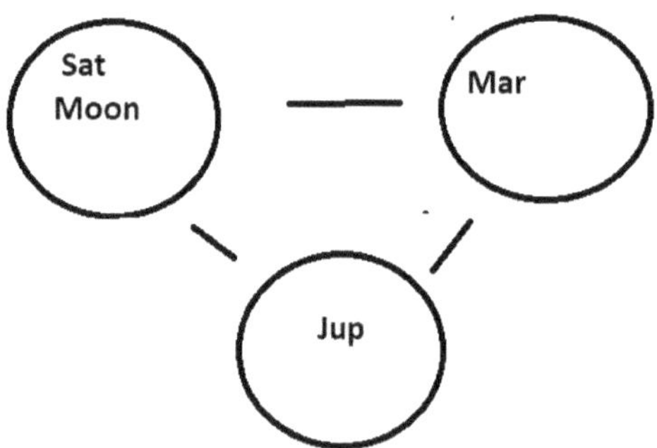

**Chart 4:** Exchange between Jupiter and Mars. CYCLIC PLANETARY EXCHANGE in the later period: Jupiter in Cancer, Moon in Leo, Sun in Aquarius, Saturn in Sagittarius.

| | | | |
|---|---|---|---|
| Mer<br>Mar<br>Jup(26) | Ven(2)<br>Jup(2)<br>Mar(15-) | Jup(14)<br>Ven(14-) | Jup(38)<br>Ven(38) |
| Sun | | | Ketu<br>Jup(50)<br>Ven(38) |
| Rahu | | | Moon<br>Jup(62)<br>Sun(63-)<br>Ven(62) |
| Sat(55-)<br>Jup(63-)<br>Moon(63-)<br>Ven(62) | Sat(54) | Sat(24) | |

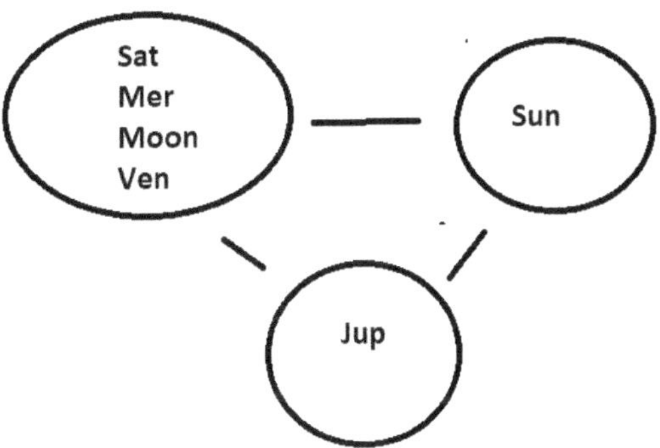

**Chart 5:** Progressional Exchange between Jupiter and Mercury and further CYCLIC PLANETARY EXCHANGE in the second round of Saturn involving Sun—Saturn-Jupiter-Moon bestow political success to the Native.

| | | | |
|---|---|---|---|
| Mer<br>Mar | | Jup<br>Rahu | |
| Sun<br>Sat(2.5) | | | |
| Ven | | | Moon |
| | Ketu | | |

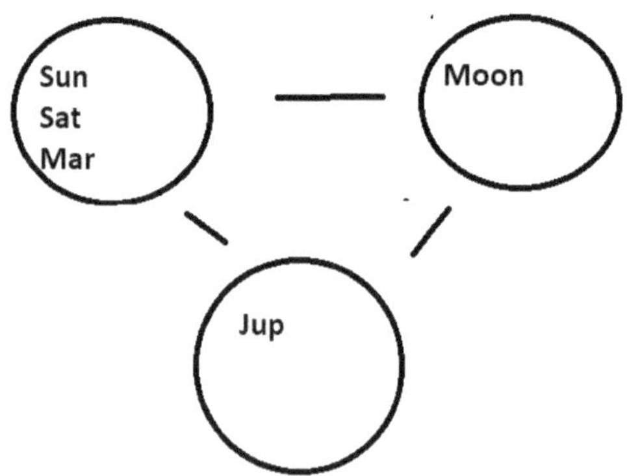

7.  DISPOSITORS: Along with their roles of Lordships, friendships, enemies, neutralities, exaltation, and debilitation, another aspect represented by planets is a phenomenon called significance, which, when considered, enhances the accuracy of prediction.

What is Significance? A planet that has Lordship of one or more houses, if placed in any sign other than its own, behaves or exhibits attributes of any planet placed in its own sign or signs. For example, consider Mars, who is the lord of Aries, being placed in Virgo and Jupiter in Aries. Due to the Rule of Significance, Mars appears to sense Jupiter and exhibits the attributes of Jupiter, although being placed in Virgo. It would amount to a Mars (Jupiter) combination operating in the horoscope. The important aspect of significance is that it explains the finer variation in all the attributes which Mars represents, thus leading to better accuracy in prediction. Not taking significance into account leads to failure in predictions. Through significance, in the example mentioned for Mars, one can describe the attributes of the younger brothers/husband (in the case of female horoscopes), the married life of the individual, the number of brothers, the quality of life led by them, the mutual relation between self and brothers, and so on. Thus, any planet placed in Aries appears to transfer its attributes to the lordship planet (Mars), irrespective of the placement of Mars in signs other than Aries and Scorpio.

In any horoscope, a planet can be placed either in its own house or in other houses. In its own house, they exhibit their intrinsic strength and direct their influence accordingly. However, if placed in houses other than their own, they transfer their influence to the lord of the house they occupy. Thus, dispositors become very important. Dispositors clarify the nature of the Karakatwas (attributes) clearly. Dispositors are useful for determining the number of wives, children, brothers and sisters, companions, their qualities, their professions, important events in their lives, timing of events, and also the nature of the profession of an individual. All the charts in this book address them.

In the chart shown, the planets whose significance the dispositor (i.e., the tenanted planet assumes) is shown in parentheses.

<table>
<tr><td>MERCURY (Ketu)</td><td>SUN(Moon)</td><td>VENUS (Saturn)</td><td>KETU</td></tr>
<tr><td rowspan="2">JUPITER (Rahu Mercury)</td><td rowspan="2"></td><td rowspan="2"></td><td>MAR(Sun)</td></tr>
<tr><td>MOON (Mar)</td></tr>
<tr><td>RAHU</td><td></td><td>SATURN (Jupiter)</td><td></td></tr>
</table>

## Few Examples Significance are given here:

**Planets in Parentheses indicate Retrogression (apparent backward motion)**

## Chart 1

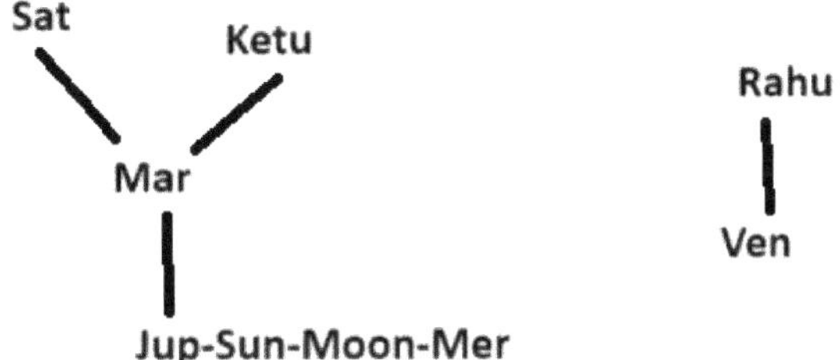

Explanation for the **Significance** tree on the right side. Jupiter is associated with Sun-Moon-Mercury in Virgo – the house of Mercury. Mars is in Pisces-the house of Jupiter. Saturn and Ketu(Moon's Node) is in the houses of Mars. Venus is in his own house Libra with Rahu(Moon's node) in Taurus.

Jupiter carrying the influence of Sun-Moon-Mercury made him a great writer and an orator. He formed a political party and became a Chief Administrator of a state.

## Chart 2

**Significances:** Saturn is in the house of Moon(Cancer); Moon with Mars is in Leo(Sun with Mercury); Sun with Mercury in Aquarius(Saturn); Venus in Pisces (Jupiter) and Ketu(Moon's node) in Libra(Venus).

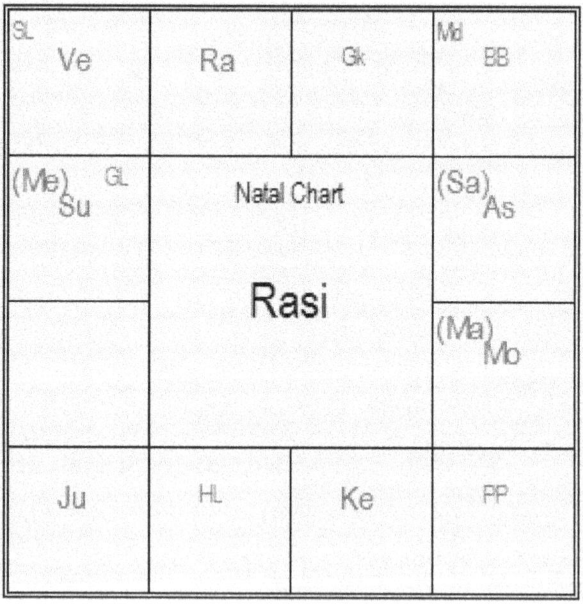

| SL<br>Ve | Ra | Gk | Md<br>BB |
|---|---|---|---|
| (Me) GL<br>Su | Natal Chart | | (Sa)<br>As |
| | Rasi | | (Ma)<br>Mo |
| Ju | HL | Ke | PP |

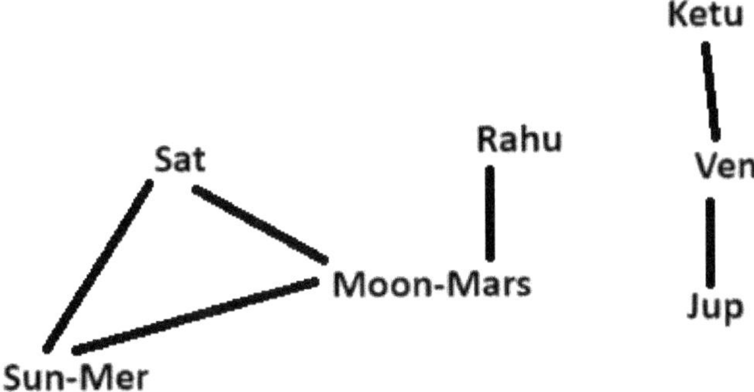


The triangular correlation between Saturn, Sun-Mercury, and Moon-Mars made this lady shine in the film industry and in politics. She headed a state as Chief Administrator, besides enduring a lot.

## Chart 3

<table>
<tr><td>Ve<br>  Ju   GL</td><td>PP    Md<br>   Ke<br>Gk</td><td></td><td></td></tr>
<tr><td>Su   Me<br>  As<br>HL</td><td colspan="2" rowspan="2">Natal Chart<br><br><br>Rasi</td><td></td></tr>
<tr><td>Mo</td><td></td></tr>
<tr><td>SL</td><td>BB<br>  Sa</td><td>Ra<br>  Ma</td><td></td></tr>
</table>

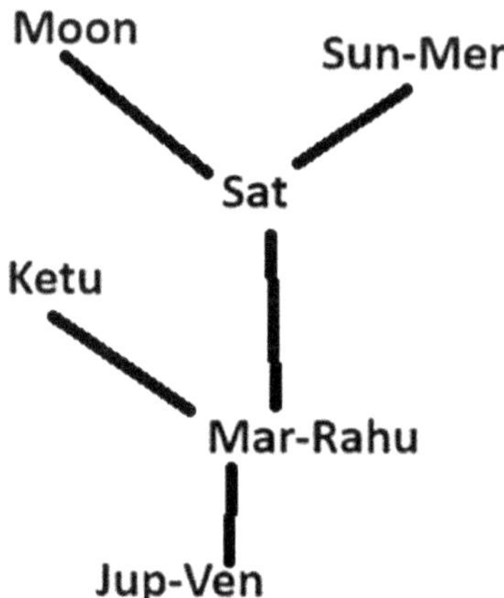



**Significances:** Jupiter with Venus in Pisces(Jupiter); Mars-Rahu(Moon's Node) in Libra(Venus); Saturn in Scorpio(Mars) and Ketu(Moon's Node) in Aries(Mars); Moon in Capricorn(Saturn and Sun with Mercury in Aquarius(Saturn).

Saturn signifying Moon and Sun-Mercury made him a head of a country, besides bestowing immense popularity. Venus signifying Mars-Rahu (Moon's node) and Mars signifying Ketu (Moon's node) led him to face a civil war for a just cause as well.

## Chart 4

| SL  PP | Ma  Me<br>Su<br>(Ve) |  | GL  Ra |
|---|---|---|---|
| | Natal Chart | | Sa  As |
| | Rasi | | Gk |
| Ju  Ke<br>Mo | | HL | Md  BB |

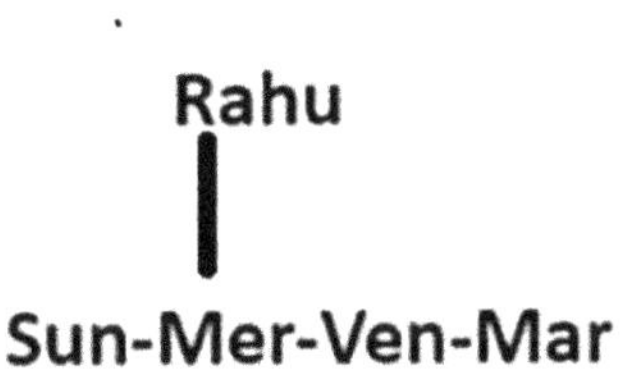

**Significances:** Jupiter with Moon and Ketu (Moon's node) in Sagittarius(Jupiter); Sun-Mercury-Venus-Mars in Aries(Mars); Saturn in Cancer(Moon); Rahu(moon's node) in Gemini(Mercury).

In Naadi astrology, the Zodiacal signs, Aries, Leo and Sagittarius are very important. Placement of planets in these signs drives the individual to act to bring great changes to the society/world. Jupiter-Moon-Ketu(Moon's node)in Sagittarius and Sun-Mercury-Venus-Mars in Aries was responsible for this individual to start a World War inflicting death, turmoil and misery in the lives of several millions of people world-wide, causing destruction and devastation in his path, which finally led to some sort of change in the perceptions of the world, besides bringing out several scientific discoveries and inventions. Astrology, as mentioned earlier, does not deal with ethical or moral questions of the actions of the individual.

## Chart 5

<table>
<tr>
<td></td>
<td>Me    HL<br>   Ma<br>BB</td>
<td>Ju    Ve<br>   Su</td>
<td>PP<br>Ke</td>
</tr>
<tr>
<td rowspan="2"></td>
<td colspan="2" rowspan="2">Natal Chart<br><br>Rasi</td>
<td>Sa</td>
</tr>
<tr>
<td>Gk<br>Mo</td>
</tr>
<tr>
<td>Ra</td>
<td></td>
<td>As</td>
<td>SL    Md<br>GL</td>
</tr>
</table>

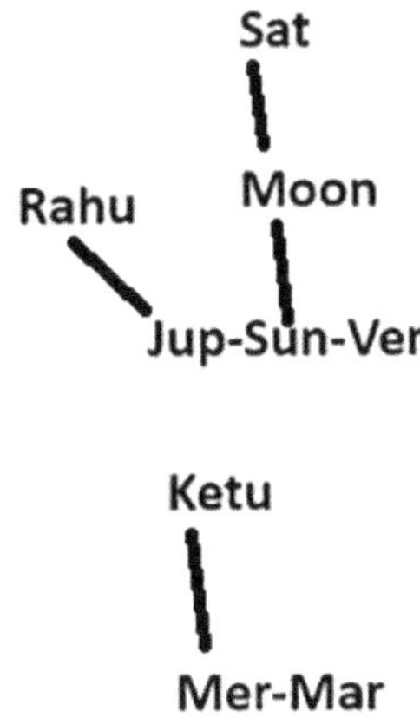


**Significances:** Jupiter-Sun-Venus in Taurus(Venus); Rahu (Moon's node) in Sagittarius(Jupiter); Moon in Leo(Sun); Saturn in Cancer(Moon); Mercury-Mars in Aries(Mars); Ketu(Moon's node) in Gemini(Mercury).

Jupiter-Sun-Venus combination along with Mars-Mercury in Aries and Moon in Leo, not only made him head of his country but also brought in great many changes to the country to make it powerful today. The same combination in Aries and Leo and Rahu(Moon's node) in Sagittarius was responsible for his violent end as well.

8.  Another type of significations, which particularly, only Naadi System of prediction employs, is the acquiring significances of the associated planets by Jupiter and Saturn. Whenever, Jupiter and Saturn are associated with other celestial bodies (including the North and South Nodes –Rahu nd Ketu), they carry their significations throughout the life of an individual, during their progressions,i.e., their traverse across the signs. Besides these two types, other two types of Significances, like that arising due to Exchange of Planets and that through probable future combinations of planets resulting from progressions of Jupiter and Saturn, also exist. We will see these things in detail in Chapter 5.

**How would one explain such significations, if there is no Science in Astrology, as the relation of planets through even Gravity amounts to only force of attraction and repulsion between planets and nothing else?**

9.  Still another important spatial influence factor in astrology, especially in Naadi Astrology, is the influence of planets in 1-5-9 positions. Planets in their positional signs at the time of birth are linked to those placed in the fifth and ninth positions. Hence, they could be grouped together as

    Planets placed in **1-5-9** positions,

    Planets placed in **2-6-10** position,

    Planets placed in 3-7-11 positions

    Planets placed in 4-8-12 positions.

    Following charts illustrate this:

## Chart 1

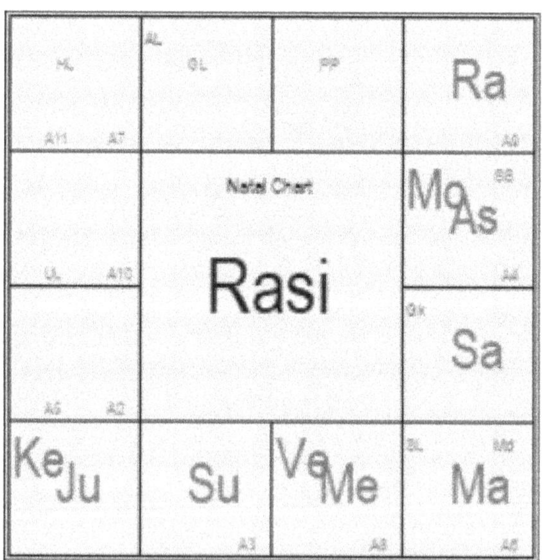


| 1-5-9 Relation | Jupiter-Ketu(Moon's node)-Saturn |
| --- | --- |
| 2-6-10 Relation | 0-0-Mars |

**3-7-11 Relation**      **0-Rahu(Moon's node)-Venus-Mercury**

**4-8-12 Relation**      **0-Moon-Sun**

## Chart 2

**1-5-9 Relation**      **Jupiter-0-0**

**2-6-10 Relation**      **Ketu(Moon's node)-0-Saturn-Mercury**

**3-7-11 Relation**      **0-Moon-Sun**

**4-8-12 Relation**      **0-Rahu(Moon's node)-Mars-Venus**

10. Time concept of Astrology is quite astounding. In this, each individual is ascribed to a unique Time Scale which is distinctly different from another individual. Each person goes through a particular pattern of events differentiated by the period for occurrences of such events. Planetary progressions like that of Jupiter and Saturn clearly help in predicting the occurrence of events, bringing out the relation between planetary progressions and the multi-faceted dimensional attributes described by Astrology. Astrology considers human beings (probably all forms of life as well)

as Space-Time entities as they describe both morphological attributes as well as the time component of the individual. Thus, Astrology takes into account the time component of each and every individual born in different times (including minute-hour-day). It is now too early to question whether our genes possess time attributes also. Our science has not advanced to that level yet. However, planetary control of our lives as a function of time is a reality in astrology.

11. Astrology considers the current planetary positions with those at birth for predictions. It is as though the planetary positions at the time of birth have created an indelible aura around the individual, probably through interactions with the genetic make-up. Hence, the current planetary positions, through the forces that they exert, influence the aforementioned 'aura' by modifying or varying it, to the extent possible, through such interactions. Predictions of major events, good or bad, are always considered with respect to the present planetary positions in comparison to those at birth for greater accuracy. Such an interaction makes a lot of sense, especially when we learn about something like 'Kirilian Images'. It is as if a dimensional distortion is created around an individual depending on the planetary positions at birth, which undergoes deformations or transformations by the influence of current planetary positions, influencing the person and events occurring in the life of an individual. The events in one's life are not only dictated by the then (present) planetary positions in his/her Natal (Birth Time) horoscope, but also by the future planetary combinations that would occur at a later date, based on the progressions of Jupiter and Saturn and that of the exchanging planets, with respect to the Natal Horoscope. Thus, not only does the 'present' dictate the 'future', but the 'future' also dictates the 'present' in Astrology.

12. Among other things, astrology deals with heredity. Earlier, Michel Gauquelin(Ref 5), a French astrologer, had proven under scientific

scrutiny that in a large number of horoscopes (charts describing positions of the planets along with the Sun and Moon at the time of an individual's birth) of parents and their children, some interesting similarities can be found in terms of certain unique planetary combinations and/or their positions. A few planets, not necessarily Jupiter and Saturn (which take 11.8613 and 29.4568 years to complete one revolution around the sun and hence can easily justify repeatability), are in either similar positions or in similar combinations. Similarities in the positions of the Sun and Moon and the other planets - Mercury, Venus, Mars, Jupiter, and Saturn - in the charts of parents and children are not coincidences. If the planets do not have a genetic correlation, this above 'inheritance' will not occur. This aspect presents a strong case for planetary influence on the GMU.

What is the probability of people with planets placed in similar positions in their horoscopes? The minimum probability will be 12 since once in 12 months planets in their progression reach the same position. However, in the case of the Moon, it is once in 27 days, for Jupiter it is once in 11.8613 years, and for Saturn, it is once in 29.458 years. Then, what will be the probability of parents and children having the same planetary positions? It will be a very low number. But it exists in astrology and in many horoscopes more than one planet or planetary combinations are similar between the horoscopes of the parents, grandparents, and the children.

Who all will exhibit similarities shown in the examples below? Only genetically related people and their offspring will show such similarities. Genetically unrelated people born under identical planetary configurations (same date of birth of parents) will show such similarities to the horoscopes of their offspring, but with a different set of planetary combinations.

Here are a few charts which show Planets that fit into THREE types of similarities:

1.   Planets in the same sign
2.   Planets in 1-5-7-9 positions
3.   Exchange positions (Before/After)
4.   Through Significance

a.   Father and Son

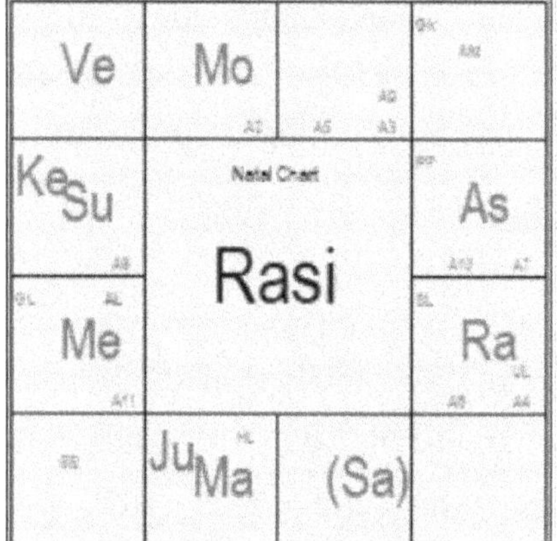


| No | Father | Son |
|----|--------|-----|
| 1 | Saturn in Libra with Moon opposite | Saturn in Libra with Moon |
| 2 | Sun in Aquarius | Sun opposite to Aquarius in Leo |
| 3 | Mercury in Capricorn and Venus in Pisces | Mercury Trine to Capricorn in Virgo and Venus Trine to Pisces in Cancer |

b.   Mother and Son

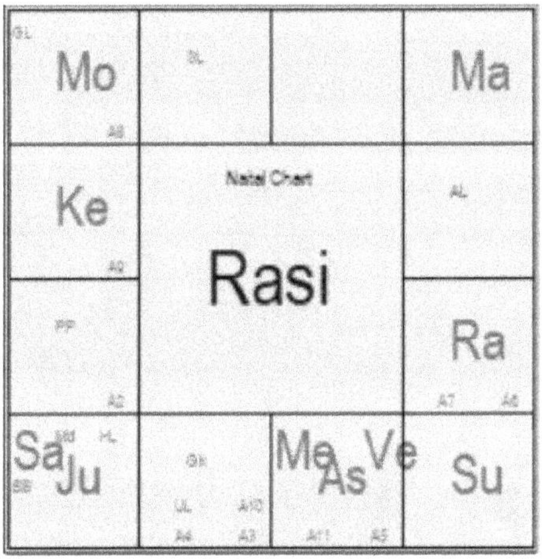

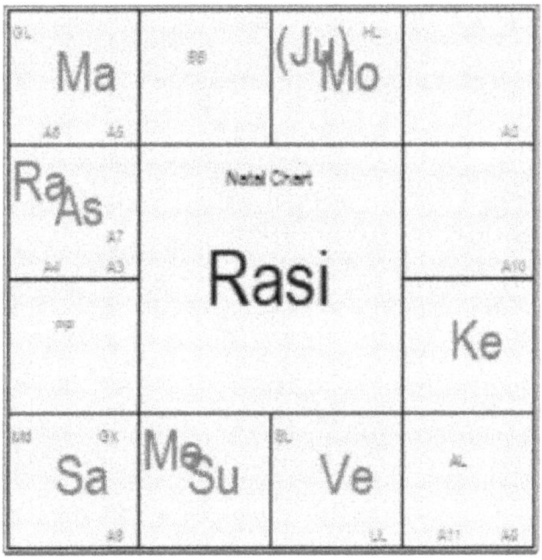

| No | Mother | Son |
|----|--------|-----|
| 1 | Venus in Libra | Venus in Libra |
| 2 | Saturn in Sagittarius | Saturn in Sagittarius |
| 3 | Mercury in Libra Signifies Mars in Gemini and Sun in Virgo | Mercury with Sun in Scorpio trine to Mars in Pisces |
| 4 | Ketu-Rahu in Aquarius and Leo | Rahu-Ketu in Aqurius and Leo |

## c.    Father and Daughter

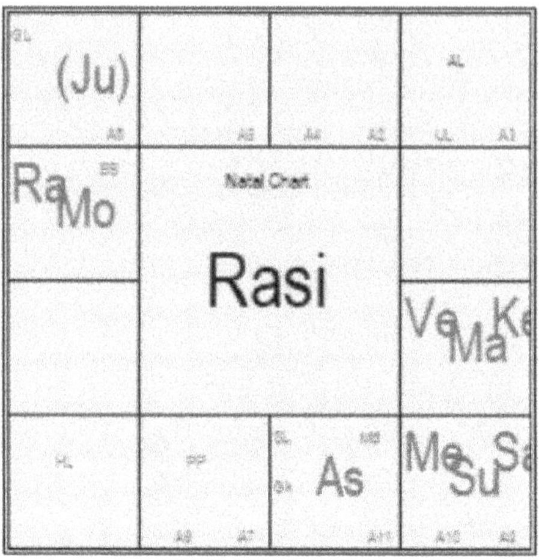

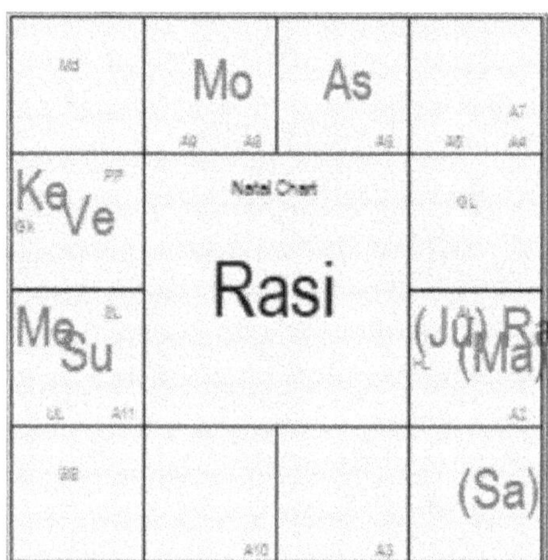

| No | Father | Daughter |
|----|--------|----------|
| 1 | Saturn in Virgo | Saturn in Virgo |
| 2 | Sun-Mercury in Virgo | Sun-Mercury in Capricorn Trine to Virgo |
| 3 | Venus –Mars in Leo | Venus in Aquarius to Mars opposite in Leo |
| 4 | Rahu-Ketu in Aquarius and Leo | Ketu-Rahu in Aqurius and Leo |

## d.  Father and Daughter

| No | Father | Daughter |
|---|---|---|
| 1 | Moon in Cancer | Moon in Cancer after Exchange |
| 2 | Sun in Scorpio | Sun in Scorpio |
| 3 | Venus in Libra Trine to Rahu in Gemini | Venus-Rahu in Sagittarius |
| 4 | Rahu-Ketu in Gemini and Sagittarius | Ketu-Rahu in Gemini and Sagittarius |

## e.   Father and Son

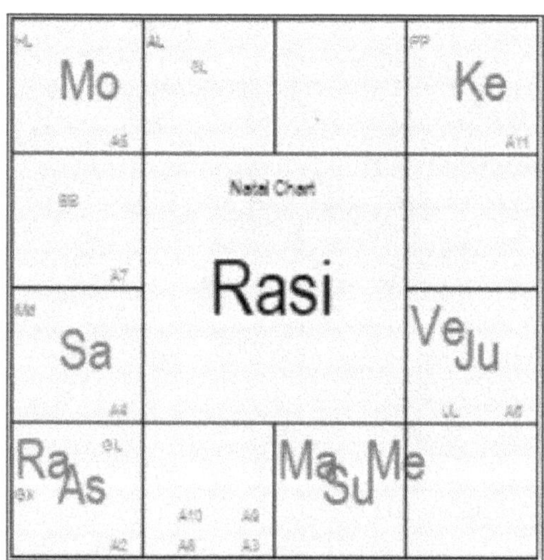

| No | Father | Son |
|----|--------|-----|
| 1 | Sun-Mars-Mercury in Libra | Sun-Mars-Mercury in Libra |
| 2 | Saturn in Capricorn | Saturn in Capricorn |
| 3 | Moon in Pisces | Moon in Pisces |

## f.  Mother and Son

| No | Mother | Son |
|---|---|---|
| 1 | Rahu-Ketu in Taurus and Scorpio | Rahu-Ketu in Taurus and Scorpio |
| 2 | Venus in Libra | Venus in Libra |
| 3 | Saturn in Aquarius | Saturn in Gemini Trine to Aquarius |

## g.    Father and Daughter

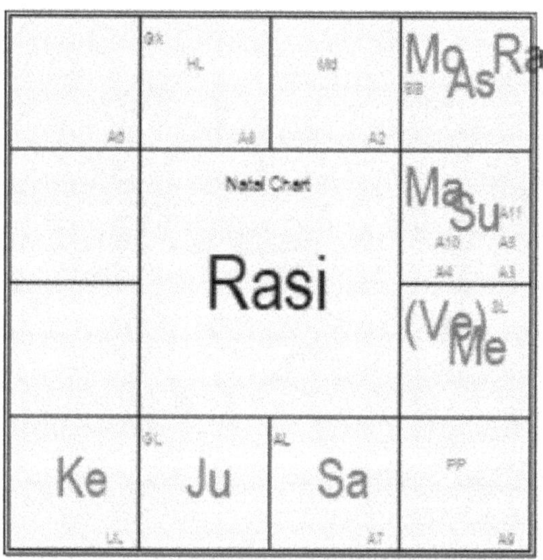

| No | Father | Daughter |
|----|--------|----------|
| 1 | Rahu-Ketu in Sagittarius-Gemini | Rahu-Ketu in Gemini-Sagittarius |
| 2 | Saturn in Libra | Saturn in Libra |
| 3 | Sun in Pisces | Sun in Cancer Trine to Pisces |

## h.  Father and Daughter

| No | Father | Daughter |
|---|---|---|
| 1 | Rahu-Ketu in Leo-Aquarius | Rahu-Ketu in Leo-Aquarius |
| 2 | Jupiter-Moon in Libra | Jupiter-Moon in Sagittarius |
| 3 | Venus in Taurus | Venus in Virgo Trine to Taurus |
| 4 | Saturn in Pisces | Saturn in Cancer Trine to Pisces |

i.   Grandfather and Grandson

| No | Grand Father | Grandson |
|---|---|---|
| 1 | Mercury in Libra signifies Mars in Virgo and Rahu in Gemini | Mercury with Mars in Virgo |
| 2 | Jupiter in Sagittarius | Jupiter in Leo Trine to Sagittarius |

j.   Grandfather and Grandson

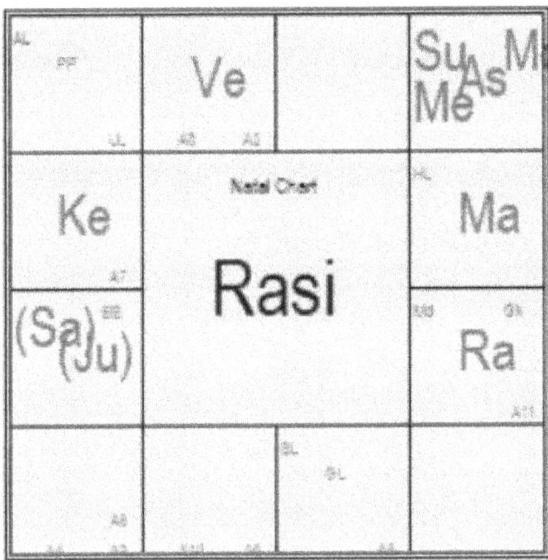

| No | Grandfather | Grandson |
|----|-------------|----------|
| 1 | Venus in Aries | Venus in Aries |
| 2 | Sun-Moon-Mercury in Taurus | Sun-Moon-Mercury in Gemini |

13. The description of an individual and events occurring in his/her life could be inferred from the horoscopes of closely related people, such as parents, spouse, children, siblings, and friends, and vice versa. Even the individual's horoscope can describe the events and

nature of people closely associated with that individual, like parents, spouse, children, siblings, and friends.

14. There are several time-tested planetary combinations in astrology. A non-believer can test himself with the following two simple tests. One such example is Chandrashtama - Moon while orbiting around the earth progresses through each sign for a period of approximately two-and-a-quarter days. When the Moon progresses to the eighth sign from the sign occupied by the Moon in one's birth chart, it is termed Ashtama Chandra or Chandrashtama period (Ashtama in Sanskrit means Eight). Every individual experiences this position once in two-and-a-quarter days in a period of 27.3207 days of the Moon's progression in his/her life several times. During this period, one or more of the following are experienced by the individual - tension, worries, unpleasant happenings, unnecessary obstructions, depressions, restlessness, temporary illness, humiliations, wasteful expenditure, suicidal tendencies, troubles in travel, etc.

Similarly, let us take the example of the most dreaded Sade Sathi period of Saturn. Sade Sathi (meaning Seven-and-a-Half in Sanskrit) is a seven-and-a-half-year period of Saturn in a period of 29.4568 years (orbital period of Saturn around the Sun), when Saturn traverses through the 12th, the 1st, and the 2nd sign to one's birth sign (where the Moon is positioned at the time of birth of the individual). Mental agony, wasteful expenses, struggle against odds, very little progress, setbacks in new ventures, diseases, losses, legal battles, a depressed outlook, additional responsibilities, difficult circumstances, fruitless travels, humiliations, uncalled-for accusations, troubles from dear and near ones, and a good period of learning and personality building are the effects one can experience during these seven-and-a-half years.

Several such features exist in astrology which are available in standard astrological books. The authors' book itself describes the effects of several such combinations. It will be beyond the scope of this book to discuss all these effects.

Chapter 3

# Newton's Gravity and Einstein's Space-Time Distortion

Astrology is, generally considered, as an empirical correlational study, dealing with the effect of planetary influences on human beings. On further examination, this empirical correlation between planets and human beings, fundamentally, deals with the influence of planets on the GMU of the human beings. The nature of this interaction is still not clear.

Scientists do not consider astrology as science, mainly because of its empiricality, lack of uniformity and rigidity. It is also true that with all the scientific knowledge, ranging from Unified Field Theory to Gravity waves, scientists and their contemporary physics cannot explain astrology. Till date, except for the attempt by Prof. Percy Seymour (Ref 6) and Michael Gaukelin (Refs 5), not much effort was made by either scientists or astrologers to prove or disprove astrology and to bring out the deep impacting science, astrology represents. It is true that in the last six centuries, our understanding of the world and universe has changed considerably. Much more paradoxes were thrown wide open after James Webb Space Telescope came up with its startling revelations. However, the scientific basis of astrology is still a far-reaching possibility for both astrologers and scientists, alike.

Ever since Gravity was considered as an improbable candidate to explain astrology, other alternatives in physics were investigated. Among the four fundamental forces – Strong Nuclear Force, Weak Nuclear Force, Electro-magnetic Force and Gravity - Gravity is the least understood. Einstein's General Relativity and Special Theory of Relativity, which are comfortable with disturbances in Space-Time continuum caused by massive objects, is silent when it comes to disturbances caused by planets on the Space-Time continuum, that could have a profound influence on the GMU of human beings.

## Newton and Gravity

Gravity baffles Scientists, even today. Newton proposed that planets are held together in their orbits around the Sun by a force of attraction balanced by the centrifugal and centripetal forces between the planets and the sun. Further, he proposed that all the objects drawn to the earth, fall on the earth and they do so by a constant acceleration with different forces acting on them depending on their masses.

Initially, it was believed that Gravity, as a force, was responsible for the astrological effects advocated. Astrologers saw a light in Newton's Theory of Gravitation. They hoped that would explain the effect of planets on the lives of human being. But that hope and euphoria was short lived. It died down once the concept of Newton's gravitational force was worked out for Earth itself.

Newton formulated an equation depicting Force of attraction or repulsion of bodies of huge masses like planets as follows:

$$F = G(M1 \times M2) / D^2$$

where F represents force of attraction or repulsion between the two huge planetary masses M1 and M2, whose force of attraction or repulsion is being examined, G the Universal Gravitational Constant and D is the distance between them. If one applies this relation to all the planets and earth we get the data mentioned in the Table 2. Although Sun is a star and Moon is a satellite of our Earth, they were considered because of the importance of these two celestial bodies (also called luminaries) in astrology.

## TABLE 2: Forces of Attraction or Repulsion between the Sun and the Planets.

| Planet | Mass in Kg | Mean Distance (m) | Distance Perigee (m) | Distance Apogee (m) | F at Apogee Newtons (Kg.m.S$^2$) | F at Perigee Newtons (Kg.m.S$^2$) |
|---|---|---|---|---|---|---|
| Sun | $1.99 \times 10^{30}$ | $1.5 \times 10^{11}$ (From Earth) | | | $3.55 \times 10^{22}$ | |
| Mercury | $3.30 \times 10^{23}$ | $5.58 \times 10^{11}$ | $4.08 \times 10^{11}$ | $7.08 \times 10^{11}$ | $7.9 \times 10^{14}$ | $2.63 \times 10^{14}$ |
| Venus | $4.88 \times 10^{24}$ | $1.08 \times 10^{11}$ | $4.14 \times 10^{10}$ | $2.58 \times 10^{11}$ | $1.14 \times 10^{18}$ | $2.93 \times 10^{16}$ |
| Moon | $7.35 \times 10^{22}$ | $3.78 \times 10^{8}$ | | | $2.05 \times 10^{20}$ | |
| Mars | $6.42 \times 10^{23}$ | $2.28 \times 10^{11}$ | $7.83 \times 10^{10}$ | $3.78 \times 10^{11}$ | $4.17 \times 10^{16}$ | $1.8 \times 10^{15}$ |
| Jupiter | $1.90 \times 10^{27}$ | $7.78 \times 10^{11}$ | $6.29 \times 10^{11}$ | $9.28 \times 10^{11}$ | $1.92 \times 10^{18}$ | $8.81 \times 10^{17}$ |
| Saturn | $5.69 \times 10^{26}$ | $1.42 \times 10^{12}$ | $1.28 \times 10^{12}$ | $1.57 \times 10^{12}$ | $1.4 \times 10^{17}$ | $9.16 \times 10^{16}$ |

Mass of Earth - $5.98 \times 10^{24}$ Kg; Universal Gravitational Constant = 6.67408x10-11 m$^3$Kg$^{-1}$S$^{-2}$. Perigee and Apogee are the closest and farthest distances respectively from earth.

(Data taken from Handbook of Chemistry and Physics, 61st Edition, CRC Press, 1980)

As can be seen from Table 1, the F values for a human being of average weight of 70 kg will be insignificant - the ratio of the mass of the person to that of the Earth is 1.17x10-23. Hence, it is criticized that a moving train passing a man standing at a distance of 100 m from the railway track will exhibit a greater effect than any planet. The F values for the two luminaries - the Sun and the Moon, are exceptionally higher than the rest of the planets. Hence, gravity as a force will never be able to explain astrology.

## Einstein's Gravity

Newton's Gravitation Theory held ground for more than two centuries. It still holds true for most of the astrophysical calculations. It underwent corrections such as the introduction of the proportionality constant in the equation described above, known as the Universal Gravitational Constant, designated as G. This was considered as the Gravitational Force exerted by the entire Universe per unit distance, unit time, and unit mass on the bodies in question.

However, the concept of gravity itself as the "force with which bodies are attracted towards the center of the Earth" underwent a change when Albert Einstein came on the scene. He is the Prophet of Science with enormous intuition. He saw what nobody saw, understood what nobody understood, and perceived what nobody perceived, in his time and beyond.

He conceived gravity as a "Distortion of Space-Time Continuum." Although all the planetary motions were explained clearly by Newton's gravitational equation, that of Mercury could not be explained. Mercury's orbit around the Sun is not fixed. It changes its orbit quite often.

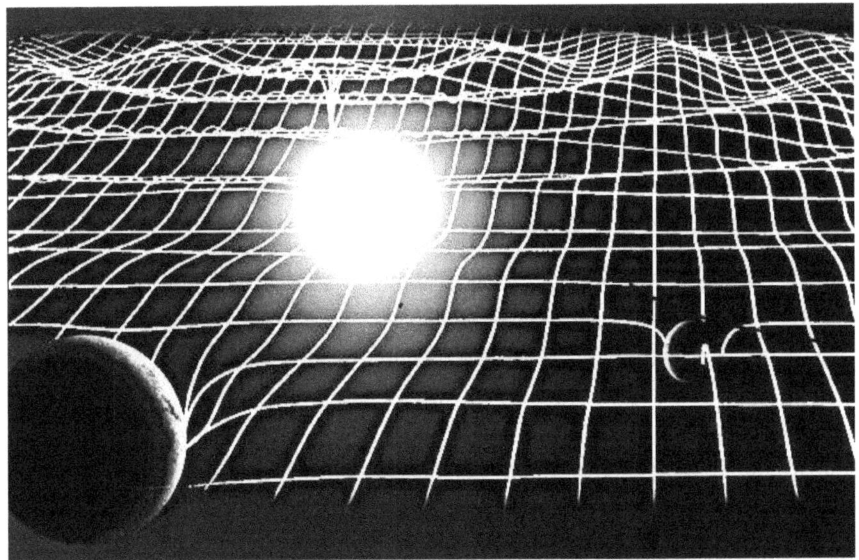

**Einstein's Space-Time Distortion (Reproduced from rk8443.blogspot.com)**

**Another view of Einstein's space-time distortion depicting Mercury's motion around the Sun (reproduced from astronomyfreaks.wordpress.com)**

Gravitational theory, especially that of the planet Mercury, posed a problem. Mercury continuously changes its orbit around the Sun, due to the enormous gravitational force of the Sun on it, which Newton's Gravitation theory failed to explain. Einstein came up with an extraordinary concept through his General Theory of Relativity. He conceived that bodies of huge masses like planets and stars, such as the Sun (and other massive components of the Universe like Black Holes,

Neutron Stars, etc.), distort the Space-Time Continuum around them, like a heavy steel ball placed in the middle of an elastic rubber sheet. He conceived a three-dimensional distortion, but we can only show this distortion in only two planes, through examples.

He then went on to explain the changing orbits of Mercury. The space around our Sun is distorted like a funnel, and a planet like Mercury will circumscribe a path in an elliptical manner if it is made to go around the Sun. This would be similar to a coin thrown along the edge of a huge funnel, slowly descending down into the funnel while traveling in an elliptical path around the axis of the funnel. Similarly, Mercury does not circumscribe a constant path around the Sun, hence changes in its orbit often occur.

The following is Albert Einstein's important field equation, which relates gravity to space-time distortion through tensors (Reproduced from Yash Quantaphypresentations, Medium)

$$G_{\mu\nu} + \Lambda g_{\mu\nu} = \frac{8\pi G}{c^4} T_{\mu\nu}$$

Where

- $G_{\mu\nu}$ is the Einstein tensor which is given as $R_{\mu\nu} - (\tfrac{1}{2} \times R g_{\mu\nu})$
- $R_{\mu\nu}$ is the Ricci curvature tensor
- $R$ is the scalar curvature
- $g_{\mu\nu}$ is the metric tensor
- $\Lambda$ is a cosmological constant
- $G$ is Newton's gravitational constant
- $c$ is the speed of light
- $T_{\mu\nu}$ is the stress-energy tensor

## Distortion in Space-Time Continuum:

Einstein saw gravity as a distortion of the space-time continuum due to the huge masses of planets. Newton's acceleration due to gravity indicates the measure of such distortion around a planet. A three-dimensional picture of such a distortion would result in planets of solid masses as 'holes' into which all bodies are attracted, leading to their 'fall'.

Gradually, scientists are getting a clear picture of Einstein's General and Special Theory of Relativity. Attempts are being made to explain all four fundamental forces - Strong Nuclear and Weak Nuclear forces, Electromagnetic force, and Gravity - in terms of General and Special Theory of Relativity, especially through the latter.

Electromagnetic force is now explained in terms of Einstein's Special Theory of Relativity. In the case of a wire carrying current, while in an inertial laboratory frame of reference, magnetic force is said to manifest; in a moving frame of reference, it is the electrical force that manifests, relatively. Hence, these two forces are now considered to be one and the same depending on our Frame of Reference, although the events are two. While equations from Ampere, Gauss, Maxwell-Heavyside, and Lorentz explain the measurement of the electrical and magnetic forces, the Special Theory of Relativity tells us what electromagnetic force is – a manifestation of one force or the other with respect to the Frame of Reference.

However, Einstein's Special Theory of Relativity fails when it comes to Quantum Mechanics and 'Singularity' of Black Holes. For example, the two Relativity theories cannot explain if the space-time around electrons, whose movement and position can only be described in termed of probabilities, is curved. They cannot also explain how an electron can be at different places at the same time. 'Singularity' of

Black holes also poses similar problems, difficult to be explained by the two Relativistic theories.

Further, Einstein's both relativistic theories are silent on the disturbances created by planets and their motions, on the Space-Time matrix. Einstein's Space-Time distortion applies to only bodies of huge masses like Sun, Quasars, Neutron Stars, Black Holes etc, besides explaining the gravity of the planets. But it fails to explain how such a distortion could affect the masses of bodies like human beings and their genetic make-up. If Planetary motions cause disturbance in the Space-Time Continuum, then definitely astrology can be explained on those lines.

In astrology, the cosmic background appears to play a significant role, with each 30° segment of the 360° space around us being dominated by different planets (called Lordships). Scientists measure the space around us in terms of Light Years, the distance travelled by light in a year, at $3 \times 10^5$ kilometer per second. In fact, what we observe when we look at the sky, is the past around us. The light that reaches us from objects in space was emitted by these objects in the past, with the time taken for light to travel equal to the distance between them and Earth in terms of light-years. This creates a time lag between the space around us and our present time on Earth. Even planets exhibit such a time lag in relation to Earth. Therefore, there are two significant time lags: one between Earth and planets, and the other between the space around us and Earth. Astrology deals with both the cosmic space around us and the planets within our vicinity.

The following table gives a rough estimate of the time lag between planets and Earth. The average distance from Aphelion and Perihelion distances of the planets with respect to the Sun was calculated. However, the time lag was evaluated from the differences in average distances between Earth and planets divided by the speed of light. The time difference is the shortest when the planet is nearer to Earth during their retrogression, and the longest when it is far away from Earth, on the 'other side of the Sun'.

Table 3 - Planetary Space-Time **Difference**[a]

| CELESTIAL BODY | DISTANCES(KM) | | AVERAGE DISTANCE (KM) | TIME DIFFERENCE (Sec)[b,c] | |
|---|---|---|---|---|---|
| | APHELION | PERIHELION | | NEAR | FAR |
| EARTH(SUN) | $1.52 \times 10^8$ | $1.47 \times 10^8$ | $1.4957 \times 10^8$ | 507 | 491 |
| MERCURY | $6.99 \times 107$ | $4.6 \times 10^7$ | $5.795 \times 10^7$ | 306 | 692 |
| VENUS | $1.09 \times 10^8$ | $1.07 \times 10^8$ | $1.0811 \times 10^8$ | 138 | 860 |
| MARS | $2.49 \times 10^8$ | $2.06 \times 10^8$ | $2.275 \times 10^8$ | 260 | 1258 |
| JUPITER | $8.16 \times 10^8$ | $7.40 \times 10^8$ | $7.778 \times 10^8$ | 2096 | 3093 |
| SATURN | $1.5 \times 10^9$ | $1.35 \times 10^9$ | $1.427 \times 10^9$ | 4261 | 5258 |

[a] From Handbook of Physics and Chemistry

[b] For Earth: (Average Distance between itself and the Sun)/(Velocity of light)

[c] For other Planets: (Diff in Average Distance between Earth and Planet)/(Speed of Light)

Speed of Light = 299793 KM/Sec

This table takes into account only the average distance from the Sun (Heliocentric Positions). Actual distances between Earth and planets (Geocentric Positions) will give a more accurate picture. However, qualitatively, the outcome is the same; there is a continuous time-lag between the positions of planets and Earth. An individual planet can show periodicity as it moves closer and farther from Earth during its revolution around the Sun. But cumulatively, if all the planets are considered at a particular time, each position of the planets with respect to Earth will never show any periodicity as each planetary position would signify a new configuration of longitude and distance to the Earth, as they keep changing their positions. Thus, with each unit passage in time, planets differ in relation to Earth, both in terms of their longitude and distance, setting up an ever-changing matrix of positions and distances.

Several concepts of Einstein's theories have been proven beyond doubt. However, Einstein's General Relativity excludes any distortion affecting human beings due to planets in our solar system. In fact, both Newton's Gravity and Einstein's relativistic theories deal only with celestial bodies of large masses. Their equations do not show any effect on objects of huge molecular assemblies of human nature.

Einstein's theory is silent when it comes to electrons in atoms and the curvature in space, if any, caused by celestial bodies of huge masses as they go round the nucleus. Human beings, in fact, exemplify a special state of matter fully composed of a huge number of molecular assemblies comprising H2O, amino acids, proteins, nucleic acids, DNA, and other smaller and bigger molecules like hormones, receptors, etc., made up of a few selected atoms like Carbon, Hydrogen, Oxygen, Nitrogen, Phosphorus, Sulfur, Zinc, Iron, Calcium, Magnesium, Sodium, Potassium, Copper, and so on. Both relativistic theories do not mention the effect of Space-Time distortion caused by celestial bodies of huge masses on these molecules. It is difficult to visualize a Universe where huge celestial masses do not exhibit any effect on atoms, molecules, and objects of human nature.

Einstein's General Theory of Relativity and Special Theory of Relativity do not have any bearing on astrology as of today. Even Einstein's concept of force disturbances traveling at the velocity of light does not appear relevant to astrology. We shall discuss this aspect in the following chapters.

Hence, a correlation between the influence of planets on one hand and the genetic material, consisting of atoms and molecules, of living beings, on the other, was considered to be only possible if it involves the quantum world of atoms and molecules.

# Chapter 4

# Probable Concepts in Physics to Explain Astrology

How can Astrology be explained scientifically? All living beings on this planet, including human beings, are mainly composed of Carbon, Hydrogen, Oxygen, Phosphorus, and Nitrogen, and to a lesser extent of Iron, Sodium, Potassium, Calcium, Magnesium, and so on. Essentially, we are all a mass of molecules assembled into human form through the process of evolution that began 3500 million years ago. Why Carbon was mainly chosen for life among all the abundant elements will take some more time for us to understand. Its tetrahedral nature, its ability to rotate plane-polarized light, and its ability to freeze into molecules of asymmetry in larger symmetric assemblies have fascinated scientists for over three centuries now. How do nucleotides in mRNA dictate what amino acid in tRNA to combine to form a protein? Which came first, RNA or Amino acids? With our present science and data, these questions cannot be answered. Without understanding these important problems, and with our present physical science concepts, the correlation between planetary influence and our GMU is not possible. Probably the existing laws of physics in the last six centuries have not been explored critically to explain a complex subject like astrology, which deals with planets of massive masses on one hand and human beings of different states of matter on the other.

*Homo sapiens* (Human beings) possess 23 pairs of chromosomes, inherited from their parents with a haploid number of 23 from each parent. Each chromosome possesses an enormous amount of genes. Each gene is made up of several hundreds/thousands of Deoxyribonucleic acid molecules called DNA. Each unit of DNA is made up of a Base, Deoxyribose sugar, and a phosphate moiety. There is only one of the four bases – Adenine, Guanine, Cytosine, and Thymine – in each Nucleotide molecule. Each gene contains these base sequences in various permutations and combinations, giving rise to a sequence that is never repeated in any other gene.

Each gene can code for a particular protein or enzyme and other such biochemicals to be synthesized in our body. From the day we are born, certain genes are 'expressed' to acquire certain attributes and certain genes are 'suppressed' not to acquire certain attributes. What governs such 'expressions' and 'repressions' is still not clear? Since, our body's tissues and organs are constituted by proteins, each gene is responsible for certain of our 'expressed' attributes – both morphological and physiological- hidden and visible.

Scientists have detected a huge amount of genetic sequences in our human genome, which do not make much sense. They are also termed as 'JUNK GENES". No genetic use has been attributed to these huge genetic materials. Their sequences do not represent any known proteins or enzymes or any other biochemical materials. It is still not known why the evolutionary process, for over 3500 million years, has left out so much of important 'unwanted' genetic material. Since we have not understood what they represent, we have termed them as 'Junk'.

All living objects are distinct from the other so called ' non-living ' matter. Living objects have different states of matter, all characterized by the magical double-helix- the DNA. It is no wonder that their role extends beyond that of mere replication and passing down of genetic information to their off-springs. Astrology definitely brings out its higher dimensional role which requires a thorough investigation. Till date only skeptical criticism has been leveled against astrology. No attempt has been made either by astrologers or scientists to understand the real role of astrology.

Here are few probable theories and concepts in physics that could explain astrology:

## Solar Magnetic Variation

Dr. Percy Seymour (Ref.6), a renowned researcher on the Magnetism of Planets and the Sun, attempted to show in 1990 that planets and human beings could be related through variations in the magnetic field of our Sun. Gravitational forces of planets could cause perturbations in the magnetic field of the Sun, leading to variations in the Solar Wind, which continuously bombard the earth. In other words, the Sun amplifies the gravitational effects of the planets in the form of magnetic field variations that are then transmitted to the earth through solar winds, affecting the fetus at the time of the child's birth by virtue of its genetic make-up. This is an interesting theory that explains astrology, although it does not explain all other aspects of astrology. For example, the effect of the background sky on the attributes of the planets cannot be explained. Spatial effects matching planets, planetary exchange, dispositor effects, planetary combinations are such astrological features that cannot be explained by magnetic variations, to name a few.

## Dark Energy –The Fifth Force:

Presently, it is established that the Universe is full of Dark Matter and Dark Energy. They are referred so, since, they cannot be 'seen'. It is estimated that both Dark Matter and Dark Energy constitutes to more than 80% of the Universal Matter and Energy.

Planets travelling across the ecliptic, could harness the Dark Energy, the extent and nature of which could depend on the planetary masses, their revolutional and rotational velocities and their ever varying distances from Earth. The differences in magnitude of the background distribution of the 'Dark Energy Flux' could explain why background Universe plays a differential role in defining the selected regions of Zodiacal Signs, their attributes and the planets that are in 'resonance' with such a background. However, this picture also does not explain several other associative features of astrology like planetary exchanges, dispositors, the role of the two nodes of Moon etc.

Astrology is a 'Life Force' that forces planetary effects to lead to CHAOS through Distraction and Obstraction leading finally to New Dimensions. The crux of astrology is this very baffling concept, which even astrologers have not bothered to pay attention to. Since Gravity cannot explain astrology and since it is also losing its status as a force according to Einstein's two Relativistic theories, astrology can only be explained by a probable force called, an 'Astrological Force' or 'Life Force'. Is it probable that Dark Matter and Dark Energy are associated with the above mentioned 'Astrological or Life Force'? With the available data and existing controversies surrounding them, Dark Energy and Dark Matter, as candidates to explain astrology has a long way to go.

## String-Super-String – Brane-M Theories:

Recently, a few theories like String, Superstring, Brane and M Theories **(References** 7-18**),** have been proposed to unify Gravity into an Unified Field Theory within one frame work, and further unifying all the four fundamental interactions. Unified Field Theory, and its related String, Superstring, Brane and M theories are still contemplating on 10-32 or more dimensional universe with four dimensions of space and time and the remaining probable hidden dimensions [7-18]. String theory requires 10 dimensions, M theory 11 dimensions and in the Bosonic String Theory the requirement is for 26 dimensions.

It is proposed in these theories, that, Plank's length particles (PLPs, 1.616x10-35 m) of one-dimension, like String, Superstring, Branes and Ms' exist all over the Universe, vibrate and is also widely distributed throughout the space, constituting a matrix. The formation of all the sub-atomic particles like electrons, protons etc, from these minute Planck's length vibrating particles are envisaged by these theories.

These theories, involve particles called Gravitons, through which Gravity effect could operates over large distances all the way to the Quantum World. Gravitons have been predicted to be mass-less particles, possessing a Spin of 2. It follows from mathematical calculations, that If Gravitons exist, they should represent at least three to six smaller dimensions. Including the three dimensions of space, they would be representing a total of 6-9 dimensions. However, the search for these smaller dimensions of Gravitons have remained elusive, even with experiments involving Large Hadron Colloiders (LDH), so far.

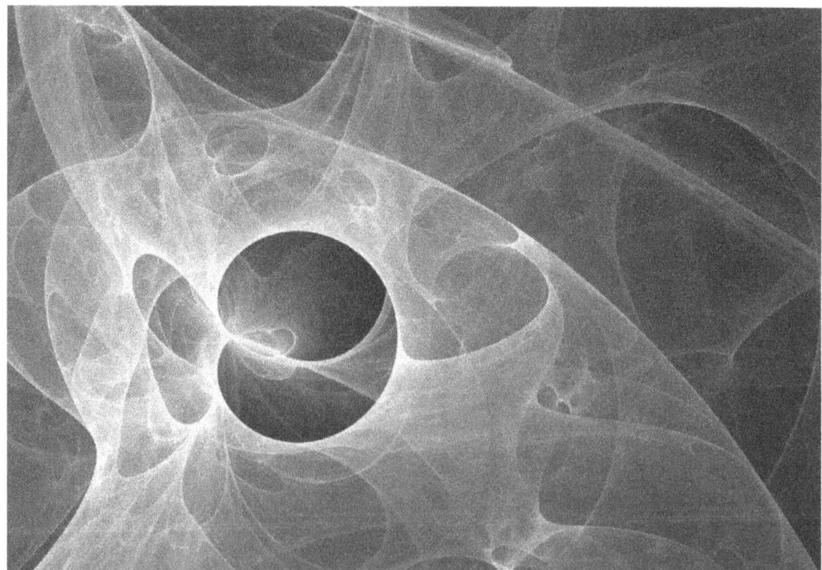

Brane: one-dimensional point-like particles of size Planck's length - 1.616x10-35 m (Reproduced from Wikipedia, The Basics of String Theory, Andrew Zimmerman Jones).

## Quantum Gravity/Loop Quantum Gravity:

Let us consider the four fundamental forces that physics advocates presently, namely – electromagnetic force, strong nuclear force, weak nuclear force, and gravity. While the Gravitational force is considered as a continuous classical field, the other three are projected as discrete quantum fields. Gravity does not have any elementary particle. But the other three exhibit a measurable unit of elementary particle. Electromagnetic force and Gravity are significant over large macroscopic scales. However, electrical and magnetic fields do not manifest in large collections of matter. Gravity so far is considered to be the dominant force over large distances. The strong nuclear force is responsible for maintaining the integrity of atomic nuclei through interactions to suppress the repulsion between sub-nuclear particles. Weak nuclear forces, on the other hand, are responsible for radioactive decay and maintaining the integrity of an atom by governing the force between electrons and the nucleus.

Scientists have attempted to integrate all four fundamental forces. However, they have been successful so far in uniting the strong nuclear force, weak nuclear force, and electromagnetism effectively. In contemporary physics, it has been difficult, so far, to link gravity to the other three forces that operate at atomic levels. Scientists are also considering a fifth force called Quantum Gravity, which can unite all four fundamental forces, although it has not attracted huge interest presently.

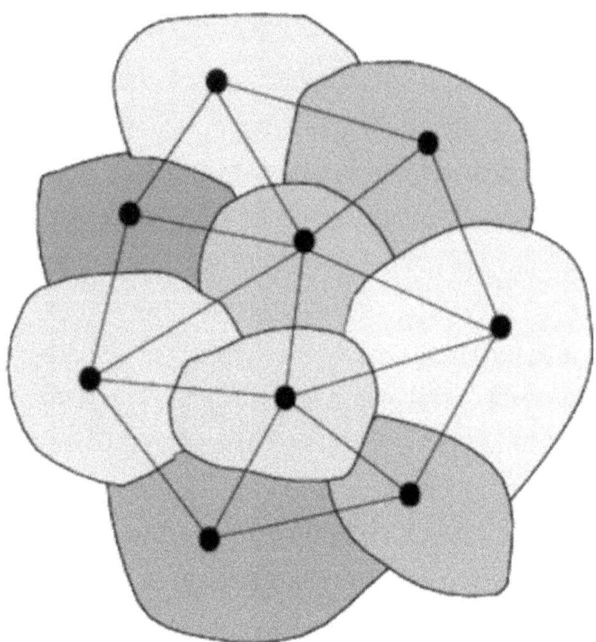

**Picture depicting Loop Quantum Gravity (Reproduced from Quora, Wikipedia)**

Gravity, as we saw in Chapter 1, is quite insignificant when planetary influences are considered on human beings. If the influence is not due to gravity, then what is it? Some physicists do not even consider gravity as a force now.

A new concept called Quantum Gravity is currently being considered by some physicists. Quantum Gravity is also proposed as a Fifth Force as an alternative to Gravity. The main objective of this consideration is to integrate Einstein's General and Special Theory of Relativity into Quantum Mechanics, for example, to explain the Space-Time curvature around electrons inside an atom.

Space-Time continuum does not support Quantum Mechanics. It depicts a continuous matrix which does not apply at the Quantum level. Hence, **Quantum Gravity or Loop Quantum Gravity** has been proposed. According to this, space is visualized to be loops of Planck's Length size

(1.16x10-35 m), which is continuous and at the same time can undergo compression and expansion at any point in space, to convey Space-Time disturbances. It is not clear whether **Loop Quantum Gravity** is also governed by the velocity of light. If it is so, it cannot explain astrological influence over huge distances.

## Lensing Effect:

The unique placement of our Moon in its near-circular orbit around Earth has baffled scientists, as no other planet in our Solar System exhibits this feature. If planets exercise their 'influence' on human beings, it could involve some sort of 'focusing'. Focusing can be twofold, one by the Planets to Earth and the other by our Moon to Earth. The Moon's orbit around our Earth defines two intersection points, North and South Nodes (termed Rahu and Ketu in Indian Astrology). These two points are attributed to both 'Divergent' and 'Convergent' attributes. While the North Node is attributed to the 'Divergent' nature, the South Node is attributed to the 'Convergent' nature, represented by the planets associated with and in relation to them. Any standard book on astrology would give more details (including the Authors' book, Ref. 2-3).

What could the planets and the two nodes focus? The following chapter attempts to explain this.

# Chapter 5

# Astrology is Quantum Entanglement

Astrology is a geocentric phenomenon. That is why, considering Earth as the center, as what astrologers did in early times, does not affect the astrological outcome. After Copernicus, we know that Earth goes round our Sun and Sun is the centre of our Solar System. But for all astrological considerations, Earth still is the centre of our astrological universe.

The Physical activities of human beings are governed by Newtonian Mechanics of force, momemtum, acceleration and so on. However, the hidden aspects associated with human beings, like qualities - beauty, intelligence, determination etc.,-, associative features concerning - Children, Wife, Parents, Grand parents, friends – and other associative features like profession, marriage, married life, profession, success, loss, gain, profit, wealth, possessions, diseases, litigations and even longevity are all dealt with by astrology. Astrology deals with these features on a comparative and probabilistic basis. Uncertainity in astrology is also associated with complexities in planetary configuration, GMU and environmental effects that throws prognostications way off the mark. But the underlying factor in predictions is probabilistic nature of these hidden features that can be 'measured' on a comparative basis only.

There is no absolute scale in the 'measurement' of all the hidden aspects which astrology deals with. In its probabilistic aspects it comes close to Quantum Mechanical concept of Heisenberg's Uncertainity principle.

Astrology definitely involves a Force –Direct or Indirect. In Astrology, planets can be envisaged to serve as 'Mediators' between the Universe, which is macroscopic and the microscopic genetic material of DNA. Newtonian mechanics deal with Universal Gravitational Constant which connects all the celestial bodies in the universe through a very small value defining force of attraction or repulsion for an unit mass, unit time and unit distance. Einstein's both General and Special Theory of Relativity, deals with a Space-Time continuum, where any disturbance should travel at the speed of light. However, the disturbance caused by Space-Time distortion by heavier masses does not speak about atoms or molecules of human beings. Just like how it cannot explain the motion of electrons around the nucleus in an atom, Einstein's General and Special Theory of relativity is silent about atoms, molecules and thereby human beings. Hence, both the relativistic theories of Einstein cannot explain astrology.

## Definition

All the concepts in physics, discussed so far do not explain all the inherent features of astrology. However, there is a most accurate and profound theory that could explain all the aspects of astrology and that is the Theory of '**Quantum Entanglement**'.

Consider two electrons with opposite spins separated over large distances, such that, it takes sometime to reach each other, even at the speed of light. It has been shown that if the spin of one electron is changed, then the spin of the other electron separated over a large distance 'changed' automatically, becoming 'opposite' to the original spin, implying that both electrons are 'linked' through a 'memory' or

'Quantum tunneling' or 'entanglement' operating at speeds greater than that of light. Einstein considered it 'Spooky' and never subscribed to this concept, since it defied his General and Special Theory of Relativity.

Quantum Entanglement "is the phenomenon of a group of particles being generated, interacting, or sharing spatial proximity in such a way that the quantum state of each particle of the group cannot be described independently of the state of the others, including when the particles are separated by a large distance', (Wikipedia). Further, it can be surmised that, 'measurement of physical properties such as position, momentum, spin and polarization performed on entangled particles can, in some cases, be found to be perfectly correlated. If the total spin, of a pair of entangled particles generated, is zero, and if one particle is found to have clockwise spin on first axis, then the spin of the other particle, measured on the same axis, is found to be anticlockwise. Any measurement of a particle's properties results in an apparent and irreversible wave function collapse of that particle and changes the original quantum state. With entangled particles, such measurements affect the entangled system as a whole" (Wikipedia).

It is for this theory that the Nobel Prize in Physics for the year 2022 was awarded to Prof. Alain Aspect, Prof. John F. Clauser, and Prof. Anton Zeilinger. The citation for their discovery quotes, "for experiments with entangled photons, establishing the violation of Bell inequalities and pioneering quantum information science."

## Astrology and Quantum Entanglement

In astrology, this concept of Quantum Entanglement, has been taken further to deal with various aspects of hidden properties of human beings, governed by human DNA molecules, to be influenced by external factors prevalent in the space through the planets, over large distances.

All our physical laws correspond to field effects which require a medium and is characterized by distance or to be more precise, space. Forces involve fields and masses (irrespective of their magnitude), be it electromagnetic or strong and weak forces or gravity. This is the major reason why astrology could not be explained by these forces that are characterized by field and mass.

Let us examine the following aspects in astrology that defy the concept of distance and mass:

1. In astrology mass does not matter – all planets exhibit more or less the same effect including the two nodes –the North and South nodes. This also includes our Moon and the massive body like the Sun.
2. Similarly, distance does not matter. While gravity or any other conceivable force varies with distance, planets, irrespective of the distances separating them, all exert the same effects.
3. Further, the region in space plays a pivotal role in astrology. Each region in space exhibits different attributes, matching that of a planet. Different signs in astrology like Aries, Taurus, Gemini, Cancer, Leo, Virgo, Libra, Scorpio, Sagittarius, Capricorn, Aquarius and Pisces have marked connotations. Astrology does not consider the vastness as empty and meaningless. The regions in space and their associated multi-dimensionalities like Galaxies, Stars, Black holes, Pulsars, Quasars, Neutron Stars, Solar Systems etc all exude an influence, irrespective of mass and distances.

    Although space appears still, our understanding of it gives us a picture that it is a dynamic system. Nothing is 'QUIET' there. There are Supernova, Solar Systems, Suns, Galaxies, Quasars, Pulsars, Neutron Stars and Black holes, most of which are in constant motion. All these constituents could be generating probably Quantum tunneling effects that could operate in the Fifth and higher dimensions to bring these effects to our door, to influence.

4.  The force or spatial influence in astrology is not an uniform' unlike the four fundamental forces in Physics- Strong and Weak Forces, Electromagnetic force and Gravity. The nature of influence exerted depends on the region in space. Although the space is divided into 12 signs of 30° each, the division is only arbitrary. However, the influence of each and every sign is different, a concept, which does not fit into any FORCE category. Further, the attributes of the influence of these signs are also matched by specific planets which are considered as lords of those signs. Thus the influence exerted links the given region in space to planets. Planets could play the role of focusing such 'spatial influences' on to the Earth.

5.  The region in space becomes the reference point for a planetary characteristic. Each planet has features that matches with the characteristics attributed to a particular region in space. A planet and a specific region in space exhibit the same attributes. This is a Quantum Entanglement effect. The space and the planet do not have any other connection except that the planet traverses a particular region across the space. A planet which is not a lord of that sign (region) exhibits modified attributes when it traverses in space across that region. It is as if, a planet has a definite configuration that matches with that of the specific spatial background and its configuration assumes different significance when it traverses across another region in space.

6.  Planets grouped under 1-5-9; 2-6-10; 3-7-11; 4-8-**12** positions play a crucial role in describing the nature of things as well as events. While the planets with Jupiter in positions 1-5-9 indicate the present state of affairs, those in 2-6-10 refer to future course of events, those in 4-8-12 to those that had happened (the past events and the background) and those in 3-7-11 to the direct influence to those in 1-5-9 positions. The planets in trine are to influence one another by modifying the attributes.

    A planet, per se, has its own attributes. Its charateristics either matches or gets modified depending on the background region it traverses. Its attributes gets further modified by the planets placed in 1-5-7-9 positions to it. To a greater extent, the planets associated with it and the planet(s) positioned in the 5th, 7$^{th}$ and 9th to it, modifies its attributes. To a much smaller extent, all the planets also affects each others attributes.

7.  Planetary Exchanges: The lords of two different signs are posited in each others sign. For example, if Jupiter is in Taurus and Venus is in Pisces, it involves exchanges between Jupiter and Venus. Jupiter, the lord of Pisces is placed in Taurus, and Venus, the lord of Taurus is placed in Pisces. Such significances imply the link between two different planets, who by virtue of their placement in each other's sign, 'Exercise' their influences over huge distances of separation, in the life of an individual

A planet in a particular sign is placed in a sign other than its own and the lord of that sign is placed in the above mentioned planet's sign, amounting to exchange. Thus, a planet is able to 'Sense' the presence of other planets placed in their signs. The said planet is somehow 'Entangled' with its own sign and the planets positioned in its sign(s). It can be said that a particular set of configuration (be it planet or space) is able to recognize some other set of planetary configuration positioned in it.

8.  A planet that has Lordship of one or more houses, if placed in any sign other than its own, not only behaves like itself by exhibiting its own attributes, but also exhibits the attributes of any planet/planets placed in its house or houses. Let us consider for example, Venus the Lord of Taurus and Libra, placed in Leo. If Moon is placed in Taurus and Mars is placed in Libra, then Venus (in Leo) exhibits the attributes of both Moon and Mars besides exhibiting its own attributes in Leo. This is termed significance in astrological parlance.

Thus a planet is able to 'Sense' planet or planets placed in its designated position in space. This is possible only if planets and space possess certain configurational similarities or otherwise '

Entanglement'. Each planet, thus becomes, a matrix of its own attributes and configuration which matches with similar spatial matrix, thereby enabling it to recognize other planetary matrices positioned in their place.

9.  Role of **Moon's** Nodes: We have been looking at the spatial and planetary correlations so far. Planets possess masses and the background space is tenanted by a whole lot of celestial bodies. However, in astrology, the two points in space -Moon's north and south nodes play a great role in predicting events in one's life. They do not act like just points in space but behaves like planets. This is baffling. What is the role played by these two points that move in an anti-clockwise direction and which exerts so much astrological influence?

    In our Solar System, all the planets orbit around the sun, more or less, in the same plane. The two intersecting points due to Moon's orbital plane and Earth's orbital plane hold, scientific significance in astrology for Entanglement effects to operate.

10. Whenever, Jupiter and Saturn are associated with other celestial bodies (including the North and South Nodes –Rahu and Ketu), they carry the significations (attributes) of the associated planets throughout the life of an individual, during their progressions.

    The dimensional configuration of Jupiter or Saturn are modified by the associative planets, thereby resulting in a much more effective matrix for astrological forces to operate.

11. Role of Jupiter and Saturn Jupiter and Saturn and their progression through the twelve signs of the zodiac are employed to predict the future course of events as and when they reach a sign which is occupied either by one or more planets or influenced by planets in 5-7-9 positions to it. Depending on the nature of planets and their attributes, the future events unfolding are predicted. In Naadi astrology Jupiter dictates, more or less, all the timing of events in one's life starting with birth, education, profession, marriage, child

birth, change of place, diseases, debts and gains, achievements and awards, general affluence or otherwise. Hence, Jupiter is referred to as 'Life Propeller'. Saturn on the other hand determines the nature, type and number of professions a person takes up in his professional career and hence, termed as Karma Karaga.

Astrology, especially Naadi Astrology, practiced in India, deals with progressions of Jupiter and Saturn with respect to an individual's Natal horoscope. According to Naadi system of predictions, Jupiter's effect is felt in one's life for a period of 11.8613 years in each sign, starting with its Natal sign, i.e. the sign of Jupiter's placement at the time of birth. The same goes for Saturn for periods of 29.4568 years. The events in one's life is not only dictated by the then (present) planetary positions in his/her Natal horoscope, but also, by the future planetary combinations that would accrue at a later date, based on the progressions of Jupiter and Saturn with respect to their natal horoscope. Thus not only 'present' dictates the 'future' but 'future' also dictates the 'present' in astrology.

12.  Earlier, Michel Gauquelin's work on **Hereditary Effects** (Ref 5) has baffled even his scientific critics. He pointed out similarities in horoscopes between Parents, Grand Parents and Children. Such similarities cannot arise without planetary influence on the GMU of the individual.

The genes which represent the attributes of the planets should be in tune with the planets at the time of birth of an individual. Then only, the full potential of planetary configurations are realized in human beings. It **is** possible that interaction between genetic material and planets uses genetic sequences in the 'Junk Gene' portion to store information **of** the planetary disturbances starting with the birth of a person and interacting with the current planetary **disturbances,** throughout an individual's life. Is it possible that these 'JUNK GENES' contain gene sequences which register and imparts some sort of 'Memory' of planetary influences to be acted upon throughout one's life? Nobody can tell currently.

13. Another type of 'Entanglement', involves the influence of closely associated people on the individual, like those of the parents, wife, children, brothers, sisters and even friends. Their horoscopes give an indication of description as well as events in the life of that very closely related individual (blood or biological relative mostly) at a particular period and vice versa.

14. Planets like Uranus, Neptune, (Pluto) are not considered in the Indian system to play much of a role. Probably, the entanglement effect by them are far too insignificant to result in any perceivable differences. The authors also do not have more data of their unequivocal involvement in the astrological effects referred in this book.

All these features are summarized in the Figure shown below.

It may sound philosophical at first but it could be true scientifically as well. A NETWORK between Space and human beings is established through planets. The whole thing appears like a Neural Network, where the planetary matrix serves as the hidden nodes, the Space as the Variables - the INPUT nodes - and comparative and probabilistic influence on human beings and their human relations as the OUTPUT nodes.

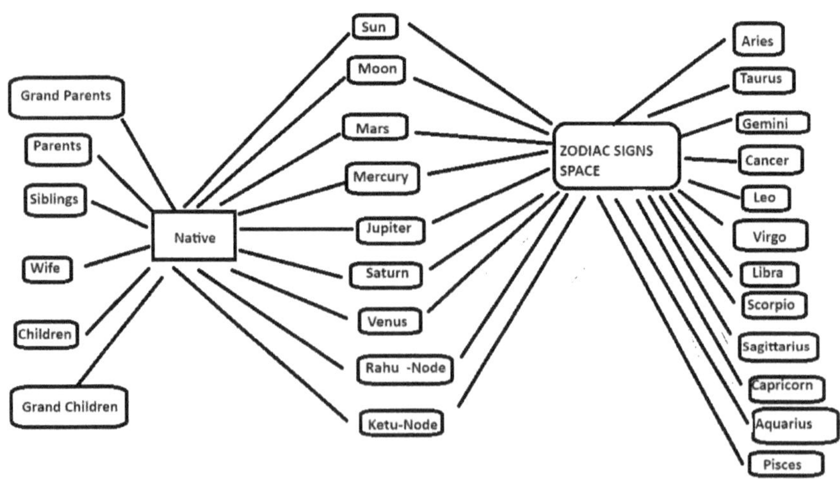


**Figure Showing all the aspects of Quantum Entanglement in Astrology**

## Atoms, Molecules and Quantum Entanglement

The various influences discussed above are not possible without 'Quantum Entanglement' like phenomenon, whose influence, imparts some sort of 'Memory' of these aspects on our genetic material to operate throughout one's life.

In the time scale of human evolution and individual lives, the background physical space shows very little observable change, irrespective of the movement of our own solar system around the center of our galaxy. The significant movements around our planet Earth are those of our own planet and those of Moon, Mercury, Venus, Mars, Jupiter and Saturn. The two nodes of Moon are also in relative motion to the Earth as they move in an anti-clockwise manner through the ecliptic path. Uranus Neptune and (Pluto) move very slowly to cause probably any measurable perceivable changes.

All the spatial objects we observe and perceive, consists of atoms including human beings. The entire cosmos is full of atoms and molecules and their assemblies integrated to form huge masses. There lies the solution to the problem. Since our understanding of such a fundamental particle like atom is not complete, we are unable to understand the complexities, which our studies throws at us, from time to time.

Our genetic material DNA, genes and chromosomes are assembly of atoms in a double helical pattern. The atoms dealt with are Carbon, Hydrogen, Oxygen, Nitrogen and Phosphorus. Between themselves, they constitute the neucleotide - Deo-Oxy ribose sugar, a phosphate moiety and one of the four bases Adenine, Guanine, Thymine, Cytosine. Structure of a DNA nucleotide is shown as:

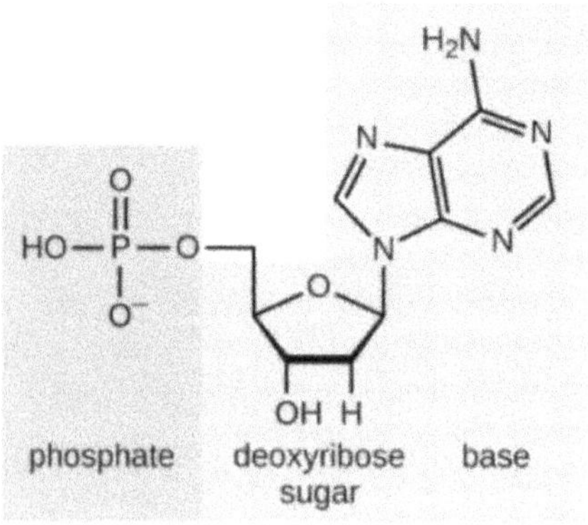
phosphate
deoxyribose
sugar
base

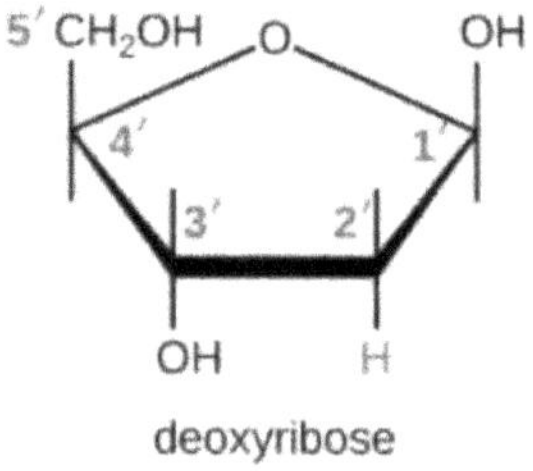
deoxyribose

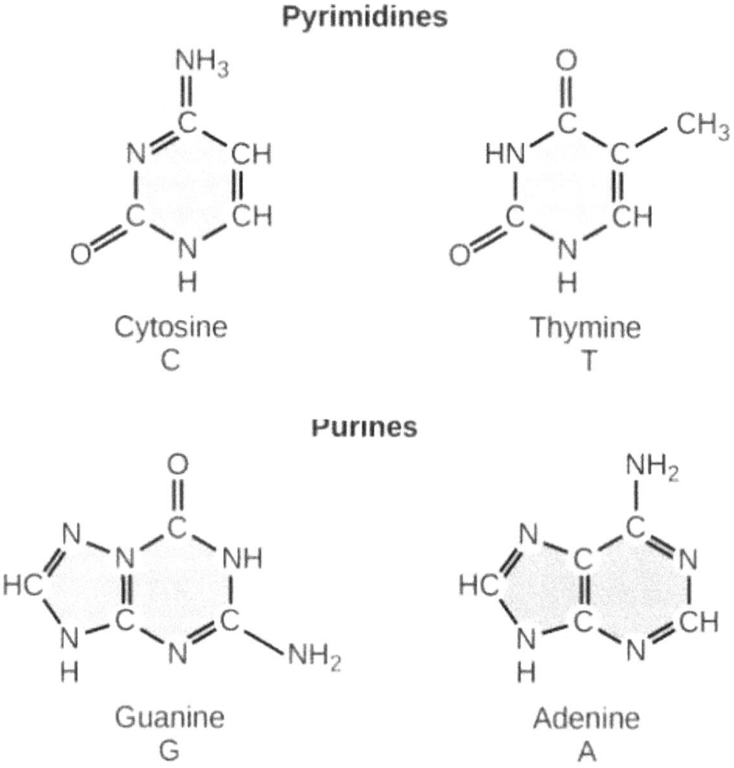
Pyrimidines
Cytosine
C
Thymine
T
Purines
Guanine
G
Adenine
A

A DNA molecule consists of two antiparallel strands of nucleotides differing only in base. A single strand of DNA consists of nucleotides bonded (linked) to one another through deoxy-ribose C3O-C5O linkages to phosphate. Both strands are held together by a series of Hydrogen-bonds between the base hydrogens. The bases in opposite helices are paired in such a manner that the double helical structure is held firmly through hydrogen bonds. Thus, Adenine pairs with only Thymine through two hydrogen bonds and Guanine with only Cytosine through three hydrogen bonds. Thus, we get a double helical structure with a repeat unit of 32-34 Å (1 Angstrom = 10-8 cm). A segment of DNA structure is shown as follows:

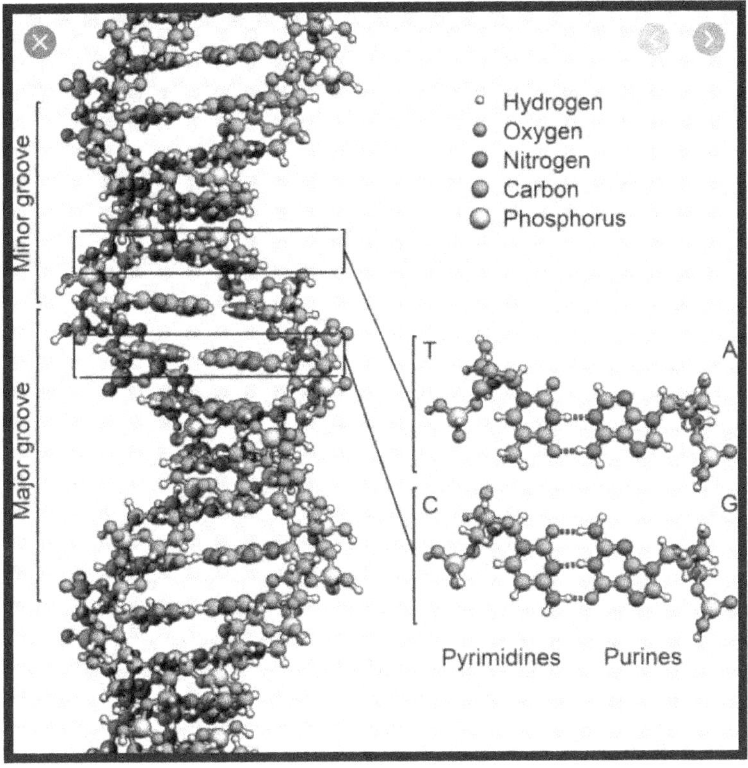


**A Segment of the DNA Molecule (Reproduced from Wikipedia)**

The Date of Birth (DOB) of an individual is an important event astrologically. To start with, the DOB is the culmination point for all the

Spatial as well as Planetary interactions transmitting the disturbances to the DNA material, which, initially, begin with the fertilized egg. Further such interactions, take place throughout the gestation period, depending on the nature of Spatial and Planetary influences, engraving their permanent signature on the foetal DNA during its development. This could create an indelible 'Memory' in the DNA material or its larger structure of the chromosomal assembly. Thus a rigid but complex network is established between the Spatial perturbations, magnified or modified by Quantum Entanglement, embedding the foetal DNA in that matrix. A Date of Birth with the specific planetary configuration carries this information only to unravel its nature as the child grows and the planets progresses. Thus, the planetary positions at the DOB, serve as a potential configuration or 'Memory' of these Planetary and Spatial interactions in the DNA templates. That is the reason, why in astrology, the planetary positions at the Date of Birth is considered to be one of primary importance, throughout an individual's life.

After the birth of the individual, the current planetary positions in an individual's life, could 'perturb' this Network (Memory) through 'modification' or 'magnification' depending on the influence from the other planetary configurations leading to all the features of astrology to manifest in an individual.

The expression or suppression of genes is initiated through 'Tweaking' this 'Memory', to affect the morphological and physiological aspects of the individual, which could eventually determine the life events of the individual. Thus, the current planetary effects could serve to activate the 'Memory' of the Genetic configuration, that had taken the entire gestation period, to be 'Put in Place'.

The manner in which children, parents and grandparents are related in astrology on one hand, and, how the individual's horoscope holds information about near blood relations like siblings and even a non-blood relation like wife/husband, clearly points to some sort of Entanglement. Thus, astrology deals with visible and hidden dimensions. So, it brings us to the question of dimensions. Besides defining morphological and physiological attributes of a human being, our human genome's ability to 'communicate' to the space around us through planets as intermediators, definitely portends a higher dimensional role to it.

DNA material is made up of atoms and molecules, whose fundamental building blocks again can be traced back to vibrating Planck's Length Particles (PLPs). If DNA is formed from basic PLPs through evolution, PLPs are the likely candidates for understanding not only astrology but also Quantum Entanglement. Their ability to 'encompass' different visible and hidden dimensions, only to manifest in DNA will be a definite possibility. DNA molecules also have the potentiality to exhibit these visible and hidden dimensions through expression and suppression of genes leading to the synthesis of proteins of specific nature required for generating such dimensions.

One way of looking at this problem is to envisage a common fundamental requirement: to explain the planetary influences on one hand and genetic material on the other, that should involve fundamental particles of an atom. Planck's length particles (PLPs - String, Super-String, Branes, and M particles) and Gravitons have been theorized to require more than three dimensions of space and one dimension of time for their vibrations.

Do the multidimensionality of Gravitons and those of vibrating String-Super String-Brane-M particles fulfill these criteria? Gravitons require 6-9 dimensions, and vibrating String-Super String-Brane-M particles of PLPs require 11-32 dimensions. Both are likely candidates for 'Entanglement,' which astrology stipulates. Quantum Entanglement, probably, also requires higher dimensions to operate. Operating at higher dimensions makes it possible for such effects to influence us through our genetic material.

However, the major 'enigma' to evoke PLPs and Gravitons for a role to explain astrology is that, they behave as particles. They possess a spin, mass, shape and size and require a medium to transmit disturbances over a long distance. They need not be the right candidates to be considered to explain astrology unless they can carry the disturbances to higher dimensions through 'Quantum Tunneling' mechanism without requirement of space and time.

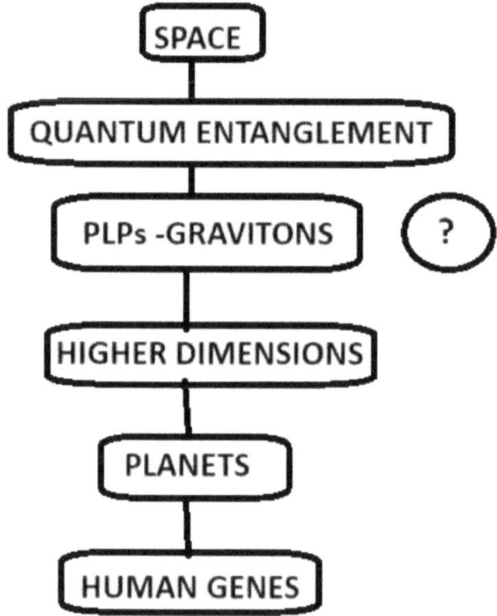


## Quantum Entanglement, Higher Dimensions and Multiverse Universes

QUANTUM ENTANGLEMENT is the best available candidate of all these to explain astrology. It does not involve a particle (?) independent of distance with an instantaneous effect. The constantly moving Sun, Moon, Mercury, Venus, Mars, Jupiter, Saturn and the two nodes of our Moon, transgressing before a region of space which do not show much change, brings forth differential influences of the whole region of their background space into focus on human beings, resulting in the Entanglement between the human beings on one hand and the vast space and its influence on the other.

Our knowledge about space itself is limited, as revealed by the findings of the James Webb Space Telescope (JWST). The Big Bang theory has come under heavy scrutiny because of JWST's findings. What is space, basically? Most people see space as an empty vast area where celestial bodies are in motion, held together by mass and distance-based attractive or repulsive forces. Einstein saw it as a Space-Time continuum, only to be distorted by bodies of heavy masses, and where disturbances could only travel at the velocity of light. Einstein attached a lot of importance to the curved surface of celestial bodies, which led him to conclude that Gravity itself is a distortion of space. He did not see Gravity as a force, as he was convinced that Gravity is a manifestation of the curvature of space around massive 'spherical' bodies. Is this a correct picture of Space?

Astrology does not see space as a continuous vast area. Celestial bodies like Stars, Planets, Pusars, Quasars, Black Holes holes etc. are treated as discrete objects, that could be interconnected with one another in a 'Quantum Tunneling ' Network. Any disturbance caused by these celestial bodies like Supernova, merging of galaxies, merging of Black Holes, and other such collisions could be transmitted to the interconnected network through 'Quantum Tunneling' effects, without invoking the concept of speed of light and need for particles.

Astrology treats human genome as a much more complex entity. It attributes a fascinating role to our human genome. When it comes to astrology, human genome reveals a much more complex role than its 'Transcriptive' and Translative abilities' to synthesise different proteins. Astrology implies a 'Communicative' role to DNA with the universe itself, as the spatial and planetary influences appears to 'control' DNA ever since a child is born. The details discussed above indicates that our human genome is capable of interacting with higher dimensions of space and planets.

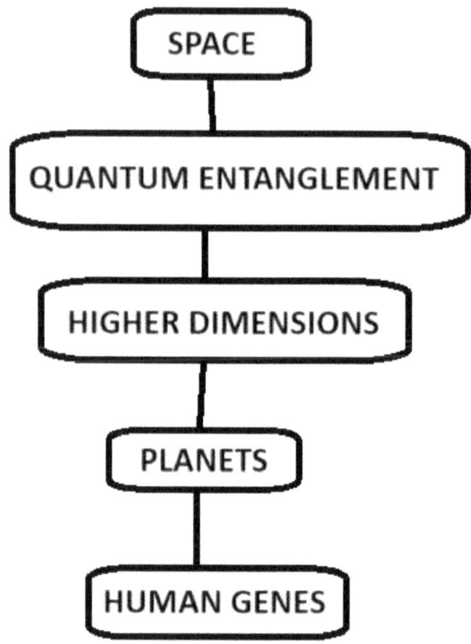


This brings one to the important conclusion. Astrology may not operate in the three dimensions of space and one dimension of time. It might require one or more multiple higher dimensions. The whole concept of astrology as discussed in the above paragraphs indicate that they do operate using hitherto unknown higher dimensions.

Already there are lot of speculations for the existence of Multiverse Universes, Parallel Universes etc. Do the twelve zodiacal signs represent 12 universes distributed all around us? Each zodiacal sign represent different astrological attributes that could be from the universe they represent. When the planets in our Solar System move across the space in the background of the signs, they could 'convey' their attributes through 'Quantum Tunneling' involving higher dimensions, to the Earth.

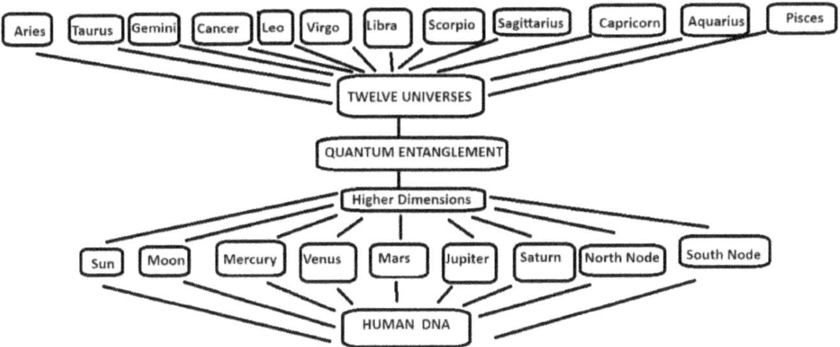


Each and every zodiacal sign represent different characteristics in astrology. The space also is not uniform around us. It is not homogenous. Each region of space has its own density of celestial objects of different nature. Perhaps they could represent even multiple universes around us capable of 'communicating' and 'controlling' us. This could also imply that the twelve signs of the zodiacal signs are actually different Universes or themselves – Multiverse entities. Thus, there could be minimum TWELVE UNIVERSES around us capable of influencing and controlling us through the propensity of higher dimensional capabilities of human genome and those of the universes.

In other words, the celestial bodies/points in our Solar System and universe could be interacting with the twelve zodiacal signs that could be twelve universes through Quantum Entanglement effects of Quantum Tunneling requiring higher dimensions. Such a scenario would open up the gateway for studying the current astrophysical findings that JWST has churned up, besides making it simple to understand our complex cosmos.

Human beings, themselves, either could be living in higher dimensions or capable of interacting with higher dimensions of Space and Planetary influences. The ecliptic plane also plays a crucial role in the above mentioned astrological influences. All the planets in our Solar System traverse in the Ecliptic plane, not above nor below. The ecliptic path followed by the planets expose the human beings to different regions of space with special emphasis on those regions of space, across which, a planet traverses. Thus, the space should possess higher dimensional components capable of interacting with higher dimensions of human genome, which human genome is capable of 'responding' in terms of planetary influences.

The varying densities of celestial bodies in each zodiacal sign are in the parallel ecliptic plane. Earth's orbital plane in the ecliptic plane is intersected by the Moon's orbital planes at the two nodes. In such a scenario, the two astrological components of Moon's North and South Nodes (termed Rahu and Ketu) could also serve as 'Quantum Tunneling points', both, receiving points of such effect of the Planets to Earth, as well as the discharging points for Earth's own effect. Hence the two nodes of Moon could assume special significance, serving as gateways to the higher dimensionally operative Quantum Entanglement to direct the influence of the zodiacal universes.

Several questions arise.

Do the nodes serve as gateways for the 'Entanglement' effects?

Do they represent higher dimensional 'Quantum Tunneling' points?

Thus, both North and South Nodes can be visulised as 'Focal points' or Quantum Tunneling ' points for both Earth, Moon and the other celestial bodies in our Solar System. Moon, mostly, could function to focus the influence of planets and background cosmos on Earth. The influence could be 'Quantum Tunneled' towards Earth by not only planets, but could be further 'Magnified' by our Moon, which is positioned exactly at the correct position.

Most fascinating aspect of Moon is its orbit around earth. It is inferred that Earth and Moon even rotate around each other, like a binary star system, with the BARY CENTER of such an orbit being within the body of Earth. It can be observed during eclipses that Moon and Earth could block Sun completely giving rise to 'Totality' by virtue of its unique position with respect to Earth. Moon's orbit is almost circular and it is exactly in the only position where it should be to block the Sun fully, when Moon comes in between Earth and Sun while traversing the nodal points, during eclipses. This is because the ratio of distances between Earth-Sun to Earth-Moon (Approx 389) is almost the same as the Ratio of sizes (diameters) of Sun/Earth to Earth/Moon (Approx 404). The probability for any satellite to assume such an orbital position is very low, if not, practically nil. No other planet in our Solar System has this kind of satellite orbiting around it.

The celestial bodies in our Solar System could transmit the INFLUENCE through a 'Quantum Entanglement process', involving 'Quantum Tunneling', whose requirements could be higher dimensions. This picture could probably be constituting the 'Astrological Force' or ' Life Force'. Depending on their nature and position encompassing the ecliptic, the role of planets could be favourable or not so favourable, such that they are designated as Lords, Friends, Enemies, Neutrals in that particular areas called signs of the Zodiac.

Human GMU could hold answer to lot of astro-physical problems. Besides being responsible for the morphological and physiological attributes of the individual, they respond to spatial and planetary influences, as indicated by astrology. This places human DNA in a unique position as it interacts with a unique 'force' of the space and planets that defies all known laws of physics till now. If this 'force' is Quantum Entanglement, then we are interacting with an INFLUENCE that require higher dimensions for operating. That brings us to an important question. Are we ourselves products of higher dimensions capable of interacting and responding to the INFLUENCES of Multiverse Universes through Quantum Entanglement effects? The answer could be yes. The entire subject of astrology points to this direction.

## More Questions and Less Answers

We have come a long way in science to understand laws of Physics and Chemistry and the human genetics. However, it appears that our understanding of these sciences are not complete yet and we still have to travel probably a long way to understand most of these things to a reasonable extent. The existing intellectual ignorance, intolerance and rejections have cost us quite a bit, as we are unable to grope with the startling realities that emerge time to time from our own studies. Recently James Web Space Telescope has thrown our whole understanding of universe into a tail spin.

Now there are further more questions to answer.

**Is it possible that Quantum Entanglement principles brings the INFLUENCE of the other universes to our door step?**

**Can astrology be the key to the understanding the interaction of other universes on us through Quantum Entanglement?**

**Is Quantum Entanglement, the Language of Nature that encompasses all the mysteries of our cosmos?**

**Who knows, how many invisible locks, such a study could open, to make our understanding of the cosmos better and complete?**

**It is high time to concentrate on Quantum Entanglement, to understand astrology.**

# Chapter 6

# Inherent Problems in Astrology

## Precession of Equinoxes

**P**recession of Equinoxes problem in astrology is a persisting problem. Both Western and Eastern Astrologers differ on this problem quite distinctly. Astronomically, 'Axial Precession' is a gravity-induced, slow, and continuous change in the orientation of a rotational axis of an astronomical body. Earth not only revolves round the Sun, but also rotates on its axis. Further, Earth also exhibits 'Nutation' at its poles, defined as the Precession of Earth's Axis, in order to balance the deviations caused by the centrifugal and centripetal forces due to rotation and revolution of our Earth. Historically called **Precession of Equinoxes**, the equinoxes move in an anti-clockwise direction along the ecliptic relative to the fixed stars. It was discovered, as early as 700 BCE and later by Hipparchus in 200 BCE. Later with accurate calculations, the period of this Precession assumed to be a variable, now stands at 25771.57 years. This corresponds to 50.3 arc seconds per year and 1 degree of precessionalmovement for every 71.6 years. However, these values are not constant as Earth's precessional speed varies constantly subscribing 49"-53" on an average, per year.

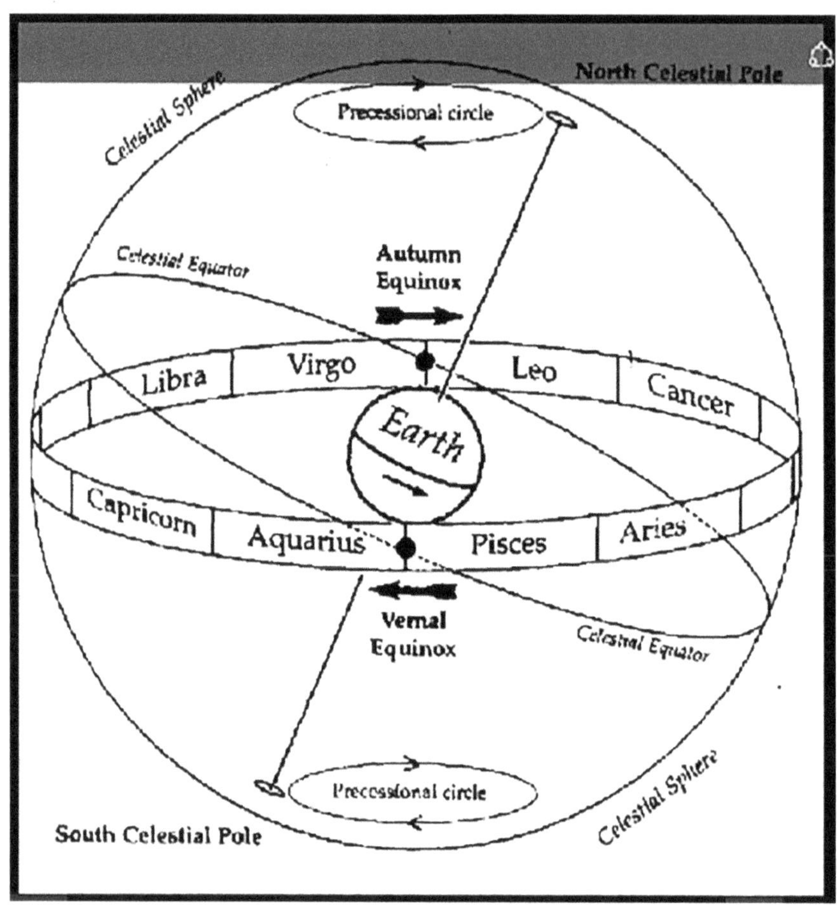


**Precession of Equinoxes along the Ecliptic (Reproduced from architexturalwatercolors.blogspot.com)**

It is not clear when astrologers started considering Precession of Equinoxes correction in defining the demarcation of zodiacal signs. Due to precession, the zodiacal sign, in front of which the Sun rises at the Vernal Equinox changes. So the demarcation point of each sign was wrongly considered to be different. The criticism posed by scientists now is that, since demarcation of signs change every year, then how can astrologers do any sort of prediction without taking the precession into account? How could such predictions made without incorporating precessional variation in casting horoscopes be called accurate?

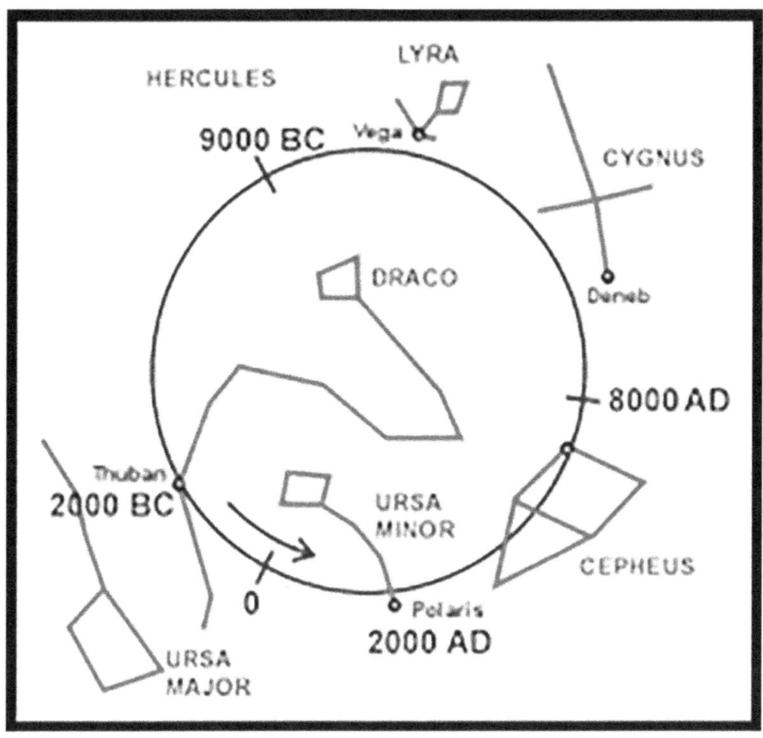


**Precession of Equinoxes (Reproduced from endgametime.wordpress.com)**

This resulted in Western astrologers, immediately responding to the criticism and implementing the precessional correction in casting the horoscopes. The demarcation points in zodiacal signs changed in Western astrology. Thus, some influential Western Astrologers did a great damage to astrologers and astrology, for Western astrology lost accuracy in predictions, thereafter. Since, no scientific basis for astrology was forthcoming in those times, such a correction, possibly could have sounded like the best thing to carry out, to overcome the criticism of scientists at those times. This has further continued, and Western Astrologers resorted to casting precessionally corrected horoscopes called 'Tropical Horoscopes'.

However, Eastern Astrologers, especially Indian astrologers, stuck to the uncorrected (for Precessional motion) horoscopes, as the Indian system always considered the concept of background FIXED stationary constellations for fixing the Zodiacal demarcation points.

## HOW TO TEST ASTROLOGY SCIENTIFICALLY?

An horoscope deals with complete configuration of planets at the time of birth and date of birth. Many, so called, scientific investigations of astrology, deals with examining correlation between a planet or a planetary combination with one or more of the other planets to a particular trait or characteristic and events. Since, many such studies showed very poor or no correlation, it was concluded by these, so called scientists, that astrology is not a science worth considering.

Why **one planet - one** correlation tests fail to **show** correlation?

Each planet exhibits particular attributes. However, a planet in a particular sign is not a simple **criteria** due to the following reasons:

1.  The attributes of the particular planet in question is modified by the sign in which it is placed.
2.  It is also modified by the associated planets along with it in the same sign.
3.  It is further modified by the planets positioned in the 5th, 7th and 9th signs to it.
4.  It is still further modified by the planets placed in sign(s) for which this particular planet is the lord and whose attributes it reflects (signifies).
5.  It is modified further, if this particular planet in question, occupying a sign, is in exchange with the lord of the sign in which it is placed, i.e. if that lord is positioned in its sign(s).

A planet in a sign **is** not **what it is** unless it is free **of all** the effects mentioned above. That is the reason why one planet one correlation will fail miserably. One can carry out this exercise. Take a planet in a sign, especially Sun sign or Moon sign and apply all the above criteria. How many cases will one find which exhibits the effect of only one planet solely? Very few cases practically. That is the reason why so much empiricality operates in astrology. A more or less statistical correlation is arrived at by astrologers based on their observation of hundreds and thousands of horoscopes. This is the reason why a situation, where, what applies to one person need not apply to another person, arises.

**Hence any test which attempts to correlate one planet to its criteria in a testing horoscope, will fail miserably and hence should not be attempted to either prove or disprove astrology.**

If planets alone could influence a person, all people born on 2nd October 1869 would have become Mahatma Gandhi. However, few similarities could be deduced in all the people born with a particular configuration of planets representing a date. Since many people are born on each and every date with a similar Planetary **Configuration,** throughout the **world** with differences in genetic make-up, the near minimum commonalities detected between them can alone prove astrology.

Naadi system offers a means of verifying this, since it deals with particular configuration of planets and does not pay much attention to the actual time of birth.

**The following study is suggested by the authors to test Astrology:**

1.  Planetary configuration of people born on any random date can be considered using the Indian system (NOT the WESTERN SYSTEM) for casting, horoscopes.
2.  Since all the planets considered are in constant motion, planetary configuration should be the same for all the people born under that particular planetary configuration. Hence, accurate duration under which the particular planetary configuration does not change, is of prime importance..

3.  Then the people should be requested to furnish details through a web site like their Date of Birth, Time of Birth and one or more of the following information. Important events that occurred in their lives along with that of their Parents, Brothers/Sisters, with details regarding individual's Education, Marriage, Children, Profession, Diseases/Disputes/Losses and Prosperity etc..

4.  The data thus collected can be analysed statistically for similarities using Artificial Intelligence techniques like Artificial Neural Network designs.

Thus, this procedure will unequivocally establish the validity of Astrology without giving any room for speculation. This experiment could be carried out by anybody with resources. One does not require astrologers to carry out this analyses. Michel Gaukelin's data failed mainly because of the approach he followed. If he had suggested this procedure, even with Western system of casting horoscopes, one could have detected quite a few similarities.

## Astrologers are a Problem NOT Astrology

Many astrologers, under the pretext of prediction, overstretch their claims, by boasting that they could give accurate predictions. Few astrologers may be good in making accurate predictions on certain aspects. Nevertheless, there is no such thing **called** accurate prediction of an entire life of an individual. This applies to Indian Astrology as well.

Then, one can ask, to what extent should **one** believe in Astrology? An horoscope for a particular day does not only pertains to the individual in question, but also to all the people born with the same configuration of planets on that date. The present estimated population of earth (in 2024) is Approximately 8.1 Billion. It comprises new born babies to old people of 100 years of age. However, as per 2012 data, the average life expectancy of a person in this world is 71 years (68.5 years for males and 73.5 years for females). How many horoscopes are required to predict the lives of all the people on this planet considering one horoscope (position/residing period of Moon in one constellation of 0-13.33°) per day? Other planetary transits to adjascent signs also occur within a day giving rise to more planetary configurations with identical planetary pattern within a given date. However, on an average let us assume that 365 or slightly more planetary configurations occur in an year, then

**No.** of **Horoscopes** = 365.25 x 100 = 36525 horoscopes

The Indian Naadi system of prediction does not give importance to the Ascendant, i.e the sign rising in the eastern horizon at the time of birth. It goes with the valid concept that, general aspects of people born under identical planetary combinations, like which occur every day during a period of unchanged planetary configuration, should be the same. Even variations, observed between people born under identical planetary combinations are recorded in Palm leaves. Thus Naadi system is quite logical and rational.

Since, only 36525 or so horoscopes are required to predict the lives and times of people of this world, on an average, how many people does a single Date of Birth represent? Assuming the population of the world to be 8.1 Billion, we can work out:

Number people, which a single horoscope could **represent** = $8.1 \times 10^9$ / 36525 = 221766

Approximately, about 222,000 people will have horoscopes with similar planetary Configurations. Will they all experience the same lives and times? The answer will be a big **No**. Even if we consider ascendants, which would amount to people born within 2 hours in any day, it would amount to:

**Number of people having the same ascendant in an identical planetary configuration** = $8.1 \times 10^9$ / (36525x12) = 18480

There are significant amount of variation in the lives, even if born under identical planetary configurations, including the Ascendant. WHY?

**Any prediction based on the individual's horoscope would be accurate only to the extent of 60% or so because two other important factors, operate which should be considered to make an accurate prediction.**

An horoscope only forms a part of a predictive tool with two other undeterminable entities- the Genetic –Make-Up (GMU) of the individual and the environment which is available to him/her, through either their own effort or forced upon them by the family and society. A planetary configuration of an horoscope representing an individual is only a road map of his/her life, outlining his/her Karma. Karma in this book, refers to all the activities of an individual carried out either for himself or his family or the government or the society **in this life only**. The authors also do not subscribe to the concept of previous birth's Karma affecting us in this life, and this life's Karma will affect us in our next birth. There is only one birth and one life and it is the one in which we are born and living presently.

 The planetary configuration does not act alone. It operates though the GMU of the individual principally impacted by the environment in which the individual is born and brought up. Unless the correlation between planets and the GMU is strong, the attributes or events indicated by the planet may not manifest fully in the individual. The inherent GMU of the person, if 'in tune' with the planetary positions, leads to, full potential of the planetary influence to manifest in that person. When such a fine tuning' is not present, then the individual and his life may not exhibit any best feature present in his horoscope, thus making him lead a mediocre life.

The first factor, the genetic make up, is further made up of two components. One is the inherent Genetic constitution of the individual and second is his/her Genealogical root. By Geneological root, we refer to the dominant planets in the individual's family as we observe similarity in planetary configuration in the horoscopes of parents and children and even those of grandparents and grandchildren. A planet placed in a similar sign in the charts of the parents and children or grandparents and grandchildren could play a big role in deciding the outcome of the attributes of the planet.

The second factor is the environment. The environment in which the individual grows up, has a profound importance in his/her life, as it can influence the extent and (to some extent) nature of things occurring in an individual's life.

Another type of 'Entanglement' involves the influence of closely associated people on the life of the individual, like those of the parents, wife, children, brothers and sisters. Their horoscopes give an indication of description of a very closely related individual as well as events occurring in life of that individual at a particular period and vice versa. As emphasized earlier, the inherent GMU of the person should be 'in tune' with the planetary positions for full potential of the planetary influence to be manifested in that person.

If the **Horoscope** of the individual could predict 60%, then the other two factors, Genetic and Environment**, could** contribute to the remaining 40% of prediction. Thus, it is important for an astrologer to consider all the above factors carefully before making an accurate prediction, which would be practically impossible. Even with ascendants, such variations can be seen between people born under identical planetary configurations.

An astrologer's role is manifold. They should be aware of the empirical nature of astrology besides understanding the environmental and genetic influence on the individual. Astrologers get carried away with the success of their predictions in some 'clear cut cases', that they themselves, fail to understand why their accuracy fail in most other cases. They pronounce with certainty some predictions which create enormous psychological aberrations in the minds of the parents and the individuals that leave a indelible mark in their mental make-up for the rest of their lives. Astrologers also 'observe' a person and give their prognostications, influenced by their own materialistic perceptions without analysing the horoscopes properly.

**Hence**, **Astrologers** should be condemned, not **Astrology**?

Many Astrologers fail miserably when it comes to prediction of major events like election results, which involves a large number of factors. It is a collective problem that cannot be assessed by just examining the chart or horoscope of a candidate. By such examination, one can only predict whether he/she will be elected and further whether he/she will occupy a high position. Here also, it is personality based prediction which is responsible for failure in predictions.

Astrologically also, it can be seen that leaders who can create more chaos are preferred to rule. Probably, this is the Law of Time, as Chaos eventually ends up in creating new dimensions.

An astrologer cannot explain an air crash or railroad train accident or a road accident. An event in astrology is not an isolated incident. It is a culmination of lot of associated factors leading to such an event and also onset of further factors associated with the affected people after the event. An accident is definitely an event involving Distraction or Obstruction in not only one's life, but also in the lives of people associated with him or her. An accident does not only terminate one's life but also cause distractions in his/her and other peoples' lives associated with the individual. A Distraction or Obstruction is governed by Moon's North and South nodes – Rahu and Ketu respectively - in conjunction with planets like Mars and Saturn.

A pillion rider dies in a motor-cycle accident, and the driver escapes unhurt. Let us say that one person has a high level of probability to die in an accident. What about the remaining people? Will their horoscopes also portend probability of death in that accident? Not necessarily. The deciding factor of number of deaths in a group accident like air, train, bus, car travel will depend on several factors.

1. The actual astrological 'culprits', i.e the individuals involved, whose horoscopes, actually pronounces danger to life at that point in life, could be very few.
2. The accident is a collective affair involving driver, other passengers, potential monetary losses incurred by the owner/insurance institutions, people whose lives will be affected by the person's death or other deformities suffered, at that point in time and afterwards and so on.

Lot of cases have to be studied to draw meaningful astrological conclusions. Such a study has never been carried out properly so far.

One comes across Sun Sign and Moon Sign predictions daily, weekly and monthly in media. There is a lot of criticism that they are not accurate. Even with 9 celestial entities, one cannot get accurate predictions. Hence, it is no wonder that these predictions based on just the progressions of Sun and Moon and few planets like Mercury, Venus and Mars cannot impart accuracy. The current planetary positions always interact with those at birth. Only those predictions that are made by taking the current planetary positions in conjunction with those at birth will be accurate. Otherwise, general predictions based **on Sun** and Moon signs **will** be **accurate,** only to the extent of 20-30%, if not still less.

 If placement of Moon decided the longitudinal duration and position of signs, with Aries (constellation Ashwini 0°) as the first sign, then the need for precessional correction for signs does not arise. The starting point of the signs will not be affected by precessional correction. Indian Astrologers never deviated from the original demarcation point of 0° in Aries, since the time of practice of astrology in India.

The precession corrected geocentric longitude of planets given by NASA ephemeris and other standard computer programs are corrected by subtracting the precessional progression from 385.33 ACE as per the Noted India Astrologer Dr BV Raman. However, this year varies from 388 BCE by Cheerio to 532 ACE by Ebenezer Burgess. Even among Indian Astrologers, this year of contention varies widely. Indian Astrology mostly uses year of correction by Lahiri's of 285 ACE and Raman's correction of 385.33ACE. Since Precessional motion is not a constant feature, as it varies from time to time, an accurate precessional ephemeris is essential for Indian astrologers to work out the longitude of the planets from NASA or other ephemeris. Thus, precessional corrections are only to arrive at the longitude of planets based on NASA or other accurate ephemeris, and not to change the signs.

Michel Gauquelin's data on **Mars Effect** and Hereditary **Effects** should be tested with the Indian System rather than with the Western System. The same chart portends different life events when examined under these two systems. It is possible that the failure of Michel Gauquelin to present a case for astrology, is probably due to use of an incorrect methodology. Hence, when tests on astrology are conducted with large number of data, the choice of the system is very important.

Since astrology deals with influence of all the celestial bodies on human lives, a few misconcepts like Sun, Moon and the two Nodes of Moon as planets are only misnomers and cannot be used to denounce Astrology.

 Also, in a continuous varying function like TIME, week-day observances like Rahu-Kala and Yamagantaka, which are given lot of importance by the people of India, do not have any meaning. This definitely is a mere superstition.

In countries like India, many astrologers offer, the so called, 'Remedies' against adverse planetary periods and positions, mainly to make money. An adverse situations arising from planetary periods and phases should be met with by change in our attitude, perceptions and behavior, rather than by rushing to the astrologers, who do not hesitate to exploit the misery of that individual, by making him spend a huge amount. A proper statistical analysis of such 'Remedies' would reveal that many such 'Remedies ' do not offer any relief. Many such statistical studies are required to correct some astrological practices to enhance the credibility of astrology.

## Anomalies in Practice of Naadi System of Astrology

It is important to mention here that many Naadi Astrologers in India misuse the system to make money. They play on the gullibility of people and resort to fraudulent practices. In the name of Naadi astrology people are taken for a ride by the so called Naadi astrologers in South India. There are lot of dubious claims in their practices.

Naadi astrology is a system of prediction based on the horoscope. It follows a method based on the longitudes of the planets, sequence of the planets, 1-5-9, 2-6-10, 3-7-11 and 4-8-12 positions of planets and so on. Jupiter and Saturn's progression through the various signs in a defined manner predicts the periods of occurrences of the events.

Because of genetic and environmental factors, the planetary configuration may not describe the person fully astrologically. So using a bundle of palm leaves they get information from the visitor who had come to them for an astrological consultation himself, by posing a series of questions to him/her. Initially they take the thumb impression of the visitor claiming that they can obtain the horoscope of the visitor from the palm leaf bundles through comparison of that thumb impression. This itself is a false claim. So far as the authors are aware, it is practically impossible to obtain the entire horoscope from the thumb impression i.e., it is difficult to arrive at the date of birth based on the thumb impression. If that is possible many difficult criminal investigations can be solved in no time because even date of births can be obtained from the horoscopes accurate to a duration of one or two days. Through series of questions, Naadi astrologers can arrive at a date of birth of the person visiting them. Therefore thumb impression is not used by these Naadi astrologers to get date of birth which is arrived at rather through the series of questions which the astrologer poses to the person. Once date of birth is obtaining then casting the horoscope for that date will do the remaining job. Their 'prediction' also becomes easy as they have already gathered enormous information about the visitor through their act of questioning and shifting the palm leaves and the bundles.

## Concluding Remarks:

Contemporary Physicists have put themselves into an Academic Straight Jacket, so as to force them to ignore the mysteries around us. Scientists should be agnostic and they should always 'think outside the box'. We may not understand all the 'puzzles' around us. But that is no reason to brand such mysteries 'unscientific' and 'nonsense'. In fact, purpose of real science should be to unravel these mysteries of nature so that we understand the universe around us for our better living. Scientists live in equations. However, these, mostly two-dimensional equations are inadequate and does not describe the universe around us. That is the reason why so many anomalies exists in science.

Our understanding can never be complete, as we tend to glorify ourselves with every 'small' breakthrough we get, thus putting a blinker on our eyes not to see any other thing around us and see only those that are ahead of us. There is also a disturbing trend in the scientific circles, which is that, every scientist appears to be working to 'find' further 'evidence' to substantiate the claims made by famous scientists who were able to explain few things rationally and quantitatively, following the established concepts up to that point in time. This prevents other scientists to even pronounce radically different concepts, for fear of being ostracized by the scientific community.

Even in science, it is only the highly 'Influential' and 'Successful' scientists, who seem to dictate what people should know and appreciate. Science is also controlled by powerful and influential Scientists and few politicians as well. Expectantly all the other scientists in the scientific levels are forced to accept this. Some do it willingly and others not so willingly. The 'punishments' met by non-acceptors are heart breaking. Their career could be destroyed completely by the influential lot. Such a kind of Scientific Borsheveism should be eradicated.

There is always an opposition to the new concepts. If logically presented, criticism can be accepted open heartedly. However, the criticism arises due to racial and other preferences. Many scientific journals, readily accept manuscripts that criticize astrology. This clearly shows that even at the scientific level people have not evolved fully to exhibit openness in dealing with controversial issues.

Physicists, Molecular Biologists, Anthropologists and Astrologers should come together to explain astrology mathematically, physically and genetically. Until that happens the enormous potential of astrology can never be harnessed by the Human Race.

# References


1. Astrology – A Science of the Quantum World, Vikram Divakar and Soundar Divakar, 2021, Notion Press, Chennai, India.

2. Language of Rahu (Naadi Rules explained), Vikram Divakar and Soundar Divakar, 2016, Notion Press, India. Available in Amazon.in and Flipkart.com

3. Rahu Naadi, Vikram Divakar and Soundar Divakar, 2024, White Falcon Publishing, Chandigarh, India.

4. Karl Marx, Das Capital, Finger Print Publishing, Vol 1,2,3, 1867-1894.

5. M. Gauquelin, Planetary Heredity, San Diego: ACS Publications, 1988.

6. P. A. H. Seymour, Astrology: The Evidence of Science (revised and extended paperback version), Great Britain: Arkana-Penguin, 1990.

7. G. Michael, J.H. Schwarz, E.Witten, Superstring theory, Cambridge University Press, 1987.

8. Vikram Divakar and Soundar Divakar, Astrology: A Key to Grand Unification, The Astrological Journal, 51(6), 60, 2009.

9. The Elegant Universe, Greene B, The Elegant Universe: Superstrings, Hidden Dimensions, and the Quest for the Ultimate Theory, Reissue edition, New York: W.W. Norton and Company, 464. ISBN 0-393- 05858-1, October 20 2003.

10. E.Witten, E, The Universe on a String, Astronomy magazine, June 2002.

11. E. Witten, Duality, Spacetime and Quantum Mechanics. Kavli Institute for Theoretical Physics, 1998, Retrieved on December 16, 2005.

12. M. J. Duff, J.T. James, T. Liu, R. Minasian, Eleven Dimensional Origin of String/String Duality: A One Loop Test Center for Theoretical Physics, Department of Physics, Texas A&M University

13. M, Kaku, Hyperspace: A Scientific Odyssey Through Parallel, Universes, Time Warps, and the Tenth Dimension. Oxford: Oxford University, Press, 384. ISBN 0-19-508514-0, April 1994.

14. L. Randall, Warped Passages: Unraveling the Mysteries of the Universe's Hidden Dimensions. New York: Ecco Press, 512 September 1 2005

15. S. Ajay, J. Wieland, The nth dimension. 2004. Retrieved on December16, 2005.

16. George Gardner, Theory of everything put to the test, tech.blorge.com, 2007-01-24.

17.  CRC Handbook of Physics and Chemistry, 61$^{st}$ Edition, CRC Press Inc., Boca Raton, USA, 1980.

18. Andrew Zimmerman Jones. "Definition of 'Brane.'" ThoughtCo, Apr. 5, 2023, thoughtco.com/brane-2699125.

19. YouTube Videos on Quantum Mechanics, Theory of Relativity, Special Theory of relativity and Astronomy.